Intelligent Systems Reference Library

Volume 219

The aim of this series is to publish a Reference Library, including novel advances and developments in all aspects of Intelligent Systems in an easily accessible and well structured form. The series includes reference works, handbooks, compendia, textbooks, well-structured monographs, dictionaries, and encyclopedias. It contains well integrated knowledge and current information in the field of Intelligent Systems. The series covers the theory, applications, and design methods of Intelligent Systems. Virtually all disciplines such as engineering, computer science, avionics, business, e-commerce, environment, healthcare, physics and life science are included. The list of topics spans all the areas of modern intelligent systems such as: Ambient intelligence, Computational intelligence, Social intelligence, Computational neuroscience, Artificial life, Virtual society, Cognitive systems, DNA and immunity-based systems, e-Learning and teaching, Human-centred computing and Machine ethics, Intelligent control, Intelligent data analysis, Knowledge-based paradigms, Knowledge management, Intelligent agents, Intelligent decision making, Intelligent network security, Interactive entertainment, Learning paradigms, Recommender systems, Robotics and Mechatronics including human-machine teaming, Self-organizing and adaptive systems, Soft computing including Neural systems, Fuzzy systems, Evolutionary computing and the Fusion of these paradigms, Perception and Vision, Web intelligence and Multimedia.

Indexed by SCOPUS, DBLP, zbMATH, SCImago.

All books published in the series are submitted for consideration in Web of Science.

More information about this series at https://link.springer.com/bookseries/8578

Bruno Apolloni

A Few Things I Know About Her

A Personally Machine Learning Inspired Approach to Understand Surrounding Nature

Springer

Bruno Apolloni
Department of Computer Science
University of Milano
Milan, Italy

ISSN 1868-4394 ISSN 1868-4408 (electronic)
Intelligent Systems Reference Library
ISBN 978-3-030-94378-3 ISBN 978-3-030-94379-0 (eBook)
https://doi.org/10.1007/978-3-030-94379-0

This Springer imprint is published by the registered company Springer Nature Switzerland AG
The registered company address is: Gewerbestrasse 11, 6330 Cham, Switzerland

Acknowledgements

After having read, either fully or in part, this book, the reader may understand my stumble with this section.

In concrete, I gratefully thank all colleagues with whom I shared thoughts and work during my many university years. In a certain sense, a scientific collaboration may prove more demanding than other kinds of partnership, as a person reveals his inner mindset and puts it in the hands of these partners, ever ready to receive any form of criticism since he knows it is honestly mutual. Formally, I'm much more cautious since I don't take for granted that some or all statements claimed in the book comply with those of my colleagues. Thus, I avoid mentioning anyone in particular, apart from Prof. Lakhmi Jain who kindly solicited me to write this book, even though I know he disagrees with some of the general ideas presented in it. That's the bliss of open-minded people.

What also embarrasses me is the fear I may have put bad mistakes somewhere in this long text. While that's a normal source of anxiety for writers of scientific texts, it is amplified for me by the wide scope of the book and, above all, by its leitmotif "issue your sentences after deep study, not simply as opinions", which, while may involuntarily sound conceited, could be cruelly contradicted by those mistakes.

So, encouraged also by friends with whom I shared my ideas, I decided to put my head in the mouth of one of my personal lions (hoping it to be tame).

Thanks for feedback.[1]

Bruno Apolloni

[1] $\alpha \& \omega$ is not a Cabala formula, it is just to note that we started the book with a lion matter and end with a lion matter as well.

Executive Summary

For starters: *the soul does not exist*. With this affirmation I no doubt lose 70% of my prospective readers—including: fundamentalists, flat-earthists, respectable people, pious types, and so forth. However, rather than seeking to provoke, I need this *incipit* to clarify that my perspective lies in bypassing the convenient soul-body dichotomy, and focus expressly on its second horn. Let's start from our brain, i.e., that huge jumble of neural cells, as the main responsible of our interaction with the surrounding Nature, and see how it may perceive, understand, and react to external stimuli. As such, this may seem to be a rather ambitious, if not unrealistic, task. Yet it is a task we face and solve in practice every day of our lives, always profiting from the experience cumulated on our body. My contribution consists simply in some strongholds that I would share with readers. Hence, I will tackle the following topics that will be the titles of as many chapters:

1. Genome + synapses, as the database and the operational root of all our actions.
2. My own personal physics, what I can know (from Sir Isaac Newton on) in search of a compatible world.
3. My favorite approach to exploiting observed data.
4. How to deal with hard-to-understand phenomena, by stressing intertwined functionalities of computation, communication, and memory activities.
5. Benefits and drawbacks of out-of-the-box solutions versus standard answers we get for free.

This short book will not represent a *Summa Teologica*, i.e., a comprehensive text where authors unveil to readers the Truth about the world (and the heavens). Rather, I will try to interpret two bases of our mental activities:*description* and *explanation*, in terms of elementary operations that take place according to a common model of our brain: namely, the weighted sum *net* and the broadcast communication through *synapses* (and their iterated application), within a diffuse broadcasting environment supported by long-term memory facilities. In my view, this leads people to navigate every day among compatible worlds instead of sticking to *the true one* (if any), where *compatibility* is another key issue of the book.

Hic sunt leones (here are lions). Folklore says ancient cartographers used these words in delimiting areas of the world they didn't visit, where lions embody the causes hindering the continuation of their explorations. Personally, I feel an analogous condition, where I know something, a few things about Her, yet touch with my hands the limits of my knowledge, and my urge to go beyond them, too, even by venturing some not-well-founded conjectures. But my mother always warned me to do not trust know-it-all people. Currently they are a plague fostered by the quick WEB search engines and the simplification conspiracy. So, in the book I will manage some professedly border-line statements, beyond which I must claim there are my personal lions.

Since the ideas so gathered concern basic issues, they should be sharable with the general reader. That's why I made every effort to write a reader-friendly text. To this aim, I did limit the amount of formulas to a minimum, and whenever possible avoid numbering formulas as well as pictures. They are exactly located under the eyes of the reader and explained in the text. By contrast, where necessary I created boxes containing somehow harder mathematics that may be skipped at a certain level of reading. In any case, at the end of each chapter I suggest further readings and related bibliography. In additional support to the reader, the text is constantly accompanied by summary notes along the margins. However, I warn the reader that the attention to be paid to the text increases chapter by chapter, where each is linked to what was discussed in the previous ones.

As an academician, I must doubly apologize: to readers for using a narrative style that is new to me; to colleagues for oversimplifying my subject matter and lacking in rigor here and there.

Thanks for reading.

Milan, Italy
October 2021

Bruno Apolloni

Figure Credits

In this section, I report the links to the sources of the images that have not been generated completely by me. I apologize in advance for some either missed or confused link, and I'm available for any correction requested by the authors of the related images.

src1 :

Lft : Wikipedia (commons/0/07/Gene.png);
Rgt : 123rf.com (57464269_l)

src2 : *inspired by*: Ospedale Pediatrico Bambin Gesù (permissionopbg.pdf)
src3 : Apolloni (Frontmatter of Algorithmic Inference in Machine LearningAdvanced Knowledge International Pty. ADELAIDE. 2003)
src4 :

Lft :Shutterstock (1032552193-huge)
Rgt : *kind permission by*: María Luisa Gandía-González (fneur-12-632036-g002.jpg)

src5 Lft : *from* `https://nwnoggin.org/resources/resource-hospitalhealth-care-outreach/`
Rgt : Wikimedia (commons/3/3e/Nerve.nida.jpg)
src6 : *from blog* `https://www.edureka.co/blog/supervised-learning/`
src7 : *from* `http://what-when-how.com/neuroscience/the-thalamus-and-cerebral-cortex-integrative -systems-part-1/`
src8 : Wikipedia (wiki/Paradosso_del_gatto_di_Schrödinger)
src9 : *shooter picture from* Depositphotos (ClayShoot_l)
src10 : *items from*: ClipArt (free-rat-clipart-36131) and Freepik (1612)
src11 : Freepik (4938)
src12 : Freepik (867)
src13 : *Top-left picture*: Shutterstock (shutterstock_619536065.eps)
src14 :

Lft : *from blog*: `http://www.edscuola.it/archivio/antologia/parlagiovane/identita/gruppo6/gruppo.htm`
Rgt : Dreamstime (dreamstime_xxl_142254934.jpg)

src15 :

Lft : Dreamstime(dreamstime_xxl_70150354.jpg)
Rgt : B. Apolloni, et Al. Social appliances for sustainable smart homes.JBSE 10(4):287–301, 2019.(figure 4)

src16 :*heads*: 123rf (9361399_l.jpg)
src17 : Depositphotos (Depositphotos_63730893_XL.jpg)
src18 : Dreamstime (dreamstime_xxl_148319296.jpg)
src19 :123rf (15713662_l.jpg)
src20 :*left* Dearmstime (dreamstime_xxl_150450480.jpg)
src21 :

frm1 : Dreamstime (dreamstime_xxl_91023130.jpg)
frm4 : Dreamstime (dreamstime_xxl_13555054.jpg)

src22 : *from* `https://www.nature.com/articles/s41586-020-2281-1?fbclid-IwAR3Z\_Q2x2Xc--1zJ9-W6v5A-VIwaeWUZGU7wWIbMB07\_ZEi8NdnaeENHHks` (figure 1)
src23 :

Lft : emphfrom blog: `https://www.genitorialcontrario.it/girotondo-della-pace/`
Rgt : Alamy (HP9GC9.jpg)

src24 : *based on* Alamy (BDCY5H.jpg)

src25 : *Rgt from public db*: `https://lapdmouse.iibi.uiowa.edu/Data/web/` (m08_Screenshot1)
src26 : *bottom right*: 123rf (43082611_l.jpg)
src27 :

Lft : ALLARRPIERSON, de collecties van de Universiteit van Amsterdam
Cntr : Wikipedia (commons.wikimedia.org/wiki/File:Prague-golem-reproduction.jpg)

src28 : *Left*: Vincos.it (Permission vincos.pdf)
src29 : *from blog*: `http://josefacaraballo.blogspot.com/2010/05/el-conejo-y-el-zorro.html`
src30 : *From blog*: `https://www.qnm.it/motori/come-consumare-meno-benzina-piccoli-trucchi-da-mettere-in-pratica-post-194975.html`
src31 : `https://www.historyofinformation.com/detail.php?entryid=2876`
src32 :

- `https://topians.wordpress.com/2015/05/13/the-allegory-of-the-cave-and-fahrenheit-451/`
- `https://pxhere.com/en/photo/1387415`

src33 : `https://acsjournals.onlinelibrary.wiley.com/doi/full/10.3322/caac.21660`
src34 :

Lft : 123rf (118812579_l_bgr.jpg)
Rgt : Alamy(FBN4YB)

src35 : Dreamstime (dreamstime_xxl_2183166)

Contents

Symbols

[]	delimiters of close intervals (those which do comprise the extremes).
()	delimiters of open intervals (those which do not comprise the extremes). They are the default delimiters when the membership of the extremes is immaterial.
{}	delimiters of a set of elements.
$\in$	denotes the membership of the left item to the right set.
$\sum_{i=1}^{n} x_i$	denotes a sum of all items in the set $\{x_1, \ldots, x_n\}$; when obvious, the extremes of the addend index are omitted.
$\prod_{i=1}^{n} x_i$	denotes a product of all items in the set $\{x_1, \ldots, x_n\}$; when obvious, the extremes of the factor index are omitted.
$\lim_{x \to x_l} f(x)$	denotes the value that $f(x)$ approaches when its argument approaches the value x_l.
∞	denotes a virtual value which is higher than any other one we can establish (it is a possible x_l in the lim notation).
dx	is an infinitely small increment of the variable x. (We may imagine it as the result of a never-ending halving of a segment whose length is whatever value of x.)
$\int_a^b f(x)dx$	denotes the integral of the function f when its argument ranges between a and b.
x's	denotes many values of x (by some notation abuse).
$\cdot$	denotes the dot product between two arrays. For $a = (a_1, \ldots, a_n)$ and $b = (b_1, \ldots, b_n)$, $a \cdot b = a_1 b_1 + \ldots, a_n b_n$

Note: For notation simplicity's sake, unless explicitly indicated I will not use different typefaces to distinguish single values from arrays and sets.

Chapter 1
Genome + Synapses

Abstract In this first chapter we start addressing the goal of understanding the surrounding Nature by starting from ourselves as a part of the Nature implementing of the goal. The core is represented by the pair genome plus synapses, to which we refer henceforth as the *golden kit*. In very basic terms, we refer to the former as a dynamic database from which to extract information to produce a brain, and to the latter as the basic components of a processing device through which to interact with Nature. Since we are not interested in a cosmogony in the book, we avoid a lot of physical and biological issues in order to focus our attention on the information content of the above database and the potentiality of the computing devices which derive from the cooperation of a multitude of neurons through the synapses that join very elementary processors so as to observe and affect Nature. The objective is to show the completeness of the golden kit in coping with the main stones of human connections with the surrounding Nature. The challenge is to share our vision with the reader by rigorously invoking the elementary operations in the kingdom of the kit. This will not lead to the conclusion that the world is easy like a talk show; rather, we will highlight the generation of a complex world from a wise sharing and composing of the results of a large ensemble of essential operators.

Far from getting involved in noble disputes on God existence (and which one), materialism, spiritualism and so on, I simply register that my mind may operate thanks to the genome, which `generates`[1] my brain and synapses that continuously update their chemico-physical features to better cope with the environment. Actually, this is not a gross sentence, like the ones I could share with a people in a bar at the aperitif time. Rather, it is an extreme synthesis of a plenty of scientific papers documenting the research progress in the last century. Let me be less synthetic, albeit hiding a lot of details.

We just need to think.

[1] Loosely speaking.

B. Apolloni, *A Few Things I Know About Her*, Intelligent Systems Reference Library 219, https://doi.org/10.1007/978-3-030-94379-0_1

1.1 Genome

If we google "genome" on the WEB we may obtain around 100 millions entries that refer to this voice, starting from a basic definition that is exactly the same (apart from a few eccentric definition in the thread of flat earth, novax, etc.). Bypassing biological details, the genome is a huge set of (around 20.000) sequences of characters of a special alphabet (made of only 4 entries[2]) that is structured in words that code the main features of our individual bodies.[3] The process from genome to features is simple in principle but very complex if we delve into details. We may consider that these sequences are stored in a database (DB) constituted by a very long (order of billions of characters) mother sequence, the DNA, organized in structured segments called *chromosomes*, from which meaningful subsegments, denoted as genes, are extracted with proper queries to be put in input to a rather sophisticated process producing proteins as the main ingredients of the biological cells constituting our body (see the two pictures below). Each cell has a DB duplicate (with a few corruptions during replication) to which it makes queries according to the cell functionality. If we consider the evolution of the species and their inner differentiations, it is a dynamic DB which changes from one individual to the next, preserving a great part of the parents' DNA.

The long-term memory.

src 1

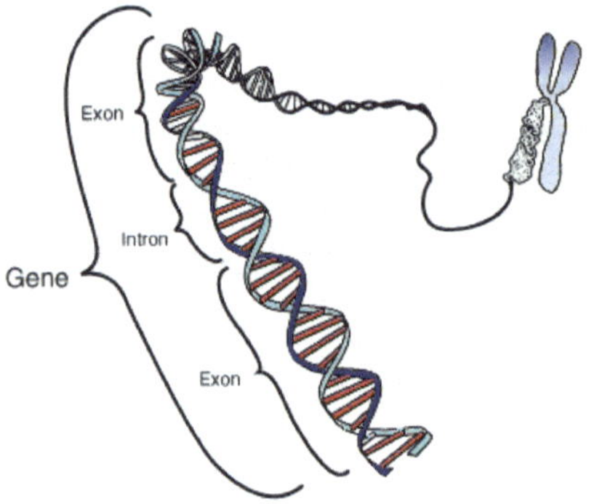

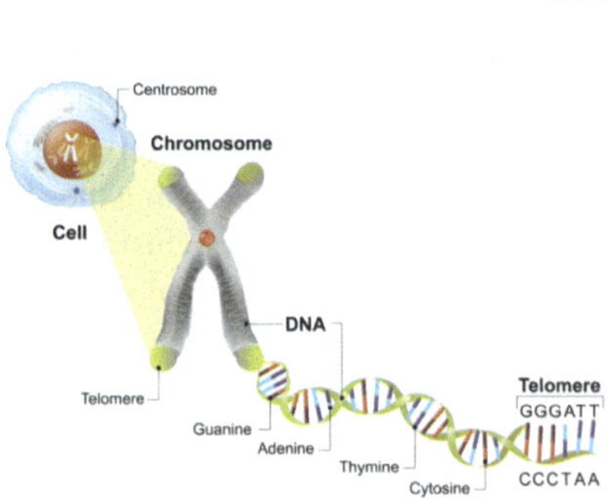

[2] Thymine (T), Adenine (A), Cytosine (C), Guanine (G).

[3] Actually, in some contexts the genome term refers not only to the the above sequence but also to the various *biological tools* such as linkers, mRNA, miRNAm, etc., which make this sequence operational.

A query tool rendered popular by the pandemic affecting our current times is the mRNA. It is a sequential structure similar to DNA, equipped with almost the same alphabet, which normally copies segments of the DNA as codes priming the production of proteins needed by our body. This happens in a protein factory whose workers are structures called ribosomes. In the case of mRNA-based vaccines, these inquirers are directly featured in the labs, so that they prime in the factory the Spike protein, which is a sort of hook through which the Covid-19 virus engages target cells of the human body. Luckily, hooks without the rest of the virus are innocuous on the one hand, and foster the production of anti-spike-antibodies, so that the protein hooks of the virus will be neutralized as well.

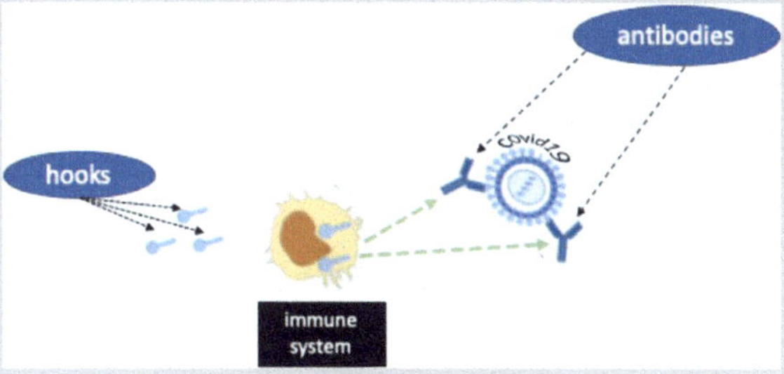

src 2

4

Ingenious messaging.

A major research effort has been devoted to this biological DB, since the discovering of DNA as the genetic material in the early twentieth-century[5] and the first sequencing of the gene coding for insuline,[6] till the whole human genome sequencing,[7] a project that has been first accomplished on a woman genome in 2004, and is currently being carried out on many other genomes.

The code for the human fabric.

Terms like DNA and chromosomes entered everyday language to denote features of our brain, our body in general, which we inherited from our ancestors and strongly characterize our habits and skills. At a pre-scientific level, it is common knowledge

---

[4] S. Clancy, W. Brown, Translation: DNA to mRNA to protein. Nat. Educ. **1**(1), 101 (2008).

[5] P. Levene, The structure of yeast nucleic acid. J. Biol. Chem. **40**(2), 415–24 (1919).

[6] A.O. Stretton, The first sequence. Fred Sanger and insulin. Genetics **162**(2), 527–32 (October 2002). PMC 1462286. PMID 12399368.

[7] "A complete human genome sequence is close" is the title of an article in the June 4, 2021 issue of Nature news (https://www.nature.com/articles/d41586-021-01506-w) which exhibits the following graph of the yearly progresses in terms of sequenced base pairs and announces that 19, 969 genes

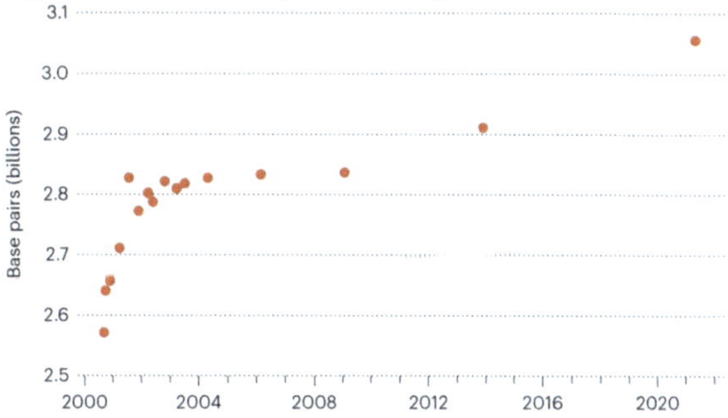

have been identified.

that the human gender is determined by the presence or absence of a particular DNA segment denoted as "chromosome Y" (actually, thanks to the presence of the SRY gene inside it). Also, the gene manipulation of plants and animals is a frequent topic in the chats of non-specialists.

In essence, nobody doubts the existence and functionality of this biological DB, albeit details are not accessible to all and the effects of those manipulations may appear scary. As for us, in this book we assume to rely on a language with a vocabulary of over 20,000 words from an alphabet of 4 characters in order to generate highly complex systems like our brains (and body in general), that prove to be different in almost each every individual populating the world. We do not get into the software to generate them, as with the lens in the picture below. While rather complex, the relative code length is definitely short in comparison to DNA length; so we may consider the software comparatively simple with respect to the variety of its output.

src 3

1.2 Synapses

Let's focus on our own magic box, the brain. It's not surprising that since the beginning of time, i.e. when man began examining his own activities and learned to think abstractly, he came up with the concept of an abstract entity, the soul, as the primary ruler of these activities. Early man denoted the heart as the locus of the soul, because when the heart "breaks" immediate death results, hence the soul leaves the body. By contrast, due to empirical relationships between state of soul and condition of brain (for instance, headache may foster non-optimal decisions),[8] the brain became

[8] Current neurological evidence shows that lesions of specific regions of the brain cause specific behavior modifications.

the preferable site for hosting the soul.[9] And in fact at a certain point people began opening the brain, not to repair damages for instance after a battle, but intentionally to investigate how it is built and how it works, getting early maps like in the figure below.

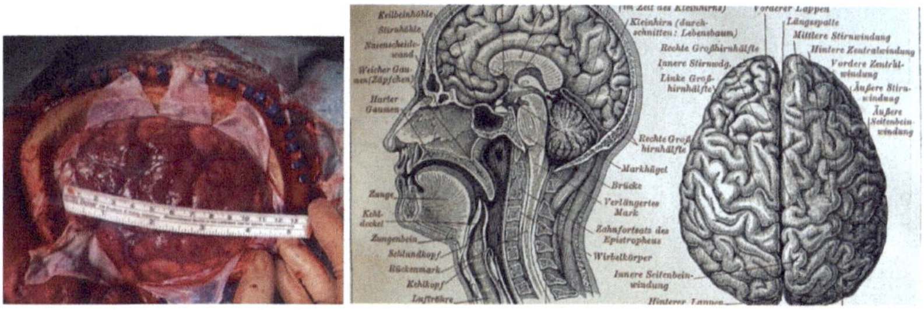

src 4

A milestone along this path consists of the Cajal brain maps obtained by staining very thin (20–100 μm thick) brain slices with silver nitrate as the contrast medium. Images like the one in the left part of the figure below reveal the basic structure of the brain to be a network of neural cells.[10] Further studies have shown that the filaments connecting the cells transmit electrochemical signals (see the right part of the figure). Namely, the fattest and longest one, called axon, is a sort of gateway propagating the signal produced by the origin cell to the others. The remaining filaments, thin and splitting like small branches of a tree, collect the signals from the axons; they are called dendrites. The synapses are the conjunctions between axons and dendrites. Investigations on how the neuron processes inbound signals and produces outbound signals are continuously in progress. Various models are under study to understand this process; however, all of them lead to a relatively simple set of ultimate operations (besides their tuning deriving from the cell-environment conditioning), so that the "intelligence" of the brain derives mainly from the communication network established by the synapses that makes available the output of a single neuron to a proper set of other neurons.

The short-term memory and computations.

[9] An even deeper question about the locus of consciousness is posed, among the others, by Mark Solms in his recent book "The Hidden Spring: A Journey to the Source of Consciousness" [13].

[10] Let's not mention the old argument about who invented this technique: Santiago Ramón y Cajal or Camillo Golgi.

src 5

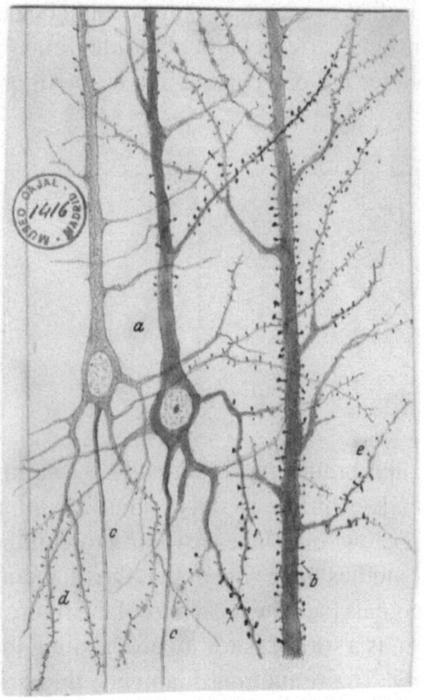

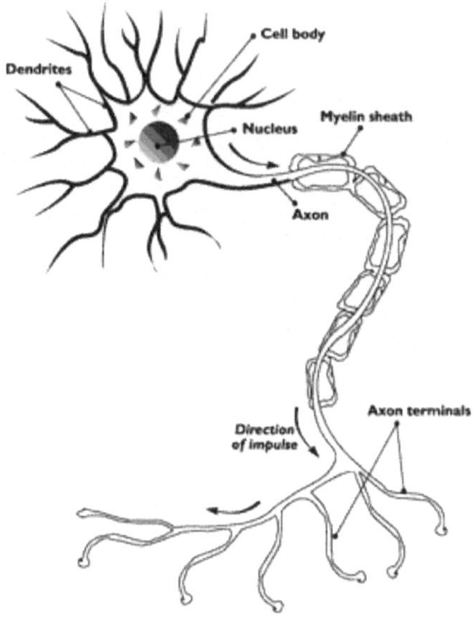

Within this topic, too, we stop at this very early step. Actually, there at least two cooperating signaling networks. Indeed, besides the one rooted on neurons, a second one exists having astrocytes (the larger portion of glia cells enveloping the brain neural network) as nodes and gap junction channels as connections [10]. Moreover, to complicate matters, micro-glia cells (the smaller glia cells) affect the operations of the former trough electrochemical interactions, so that researchers are used to model the synapses as a quad-partite junction where they converge presynaptic neurons, postsynaptic neurons and astrocytes, plus a diffuse interaction with micro-glia [11] (see Sect. 3.5 for an analogous model for artificial neurons.). Actually, brain study is a vast and wonderful discipline, but is out of the scope of the book and my personal lions are wandering on the border. Here we have just outlined some computational aspects that introduce the next section.

1.3 The Power of Artificial Neural Networks

Like with the flight of an airplane (where wings are static), which is different from that of a bird (where wings flap), an efficient emulation of brain neural networks is represented by artificial neural networks. Here the axon + synapse + dendrite is replaced by the arrowed edge of a network connecting nodes in the role of (artificial) neurons. Thus, the companion of the above Cajal map is an Artificial Neural Network (ANN) which looks like in the following picture, where arrows show the direction of the flow of the signals, let's call them *data*, circles denote the body of the artificial

Brain emulation

neuron, the neuron for short. Boxes replacing circles in the nodes represent possible variants of the canonical neuron.

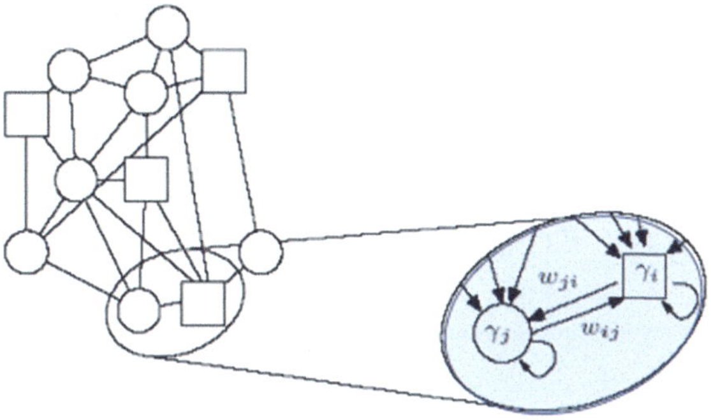

The complete unfolding of the brain's neural network is tackled generally by chunks, since the overall number of neurons in our brain is in the order of ninety billion and the number of connections between them is about ten thousand times greater.

Huge number of neurons in a real brain.

Small-scale emulation:

architecture

The companion artificial chunks are schematized through a few models, such as the Multilayer Perceptron, Recurrent Neural Networks, Convolution Neural Networks, and so on, where the differences lie in the lay-out of the edges and in the elementary computations carried out by the neurons. A key distinction in the layouts is if they:

- constitute oriented graphs, as in the left part of the next figure, so that input-output association is static, like the result of a sum of two numbers, or
- comprise loops, as in the right part of the figure, so that the output in correspondence of a given input may change with time, depending on the elapsed data loops, like with a deep meditation in our mind in search of a hard problem solution.

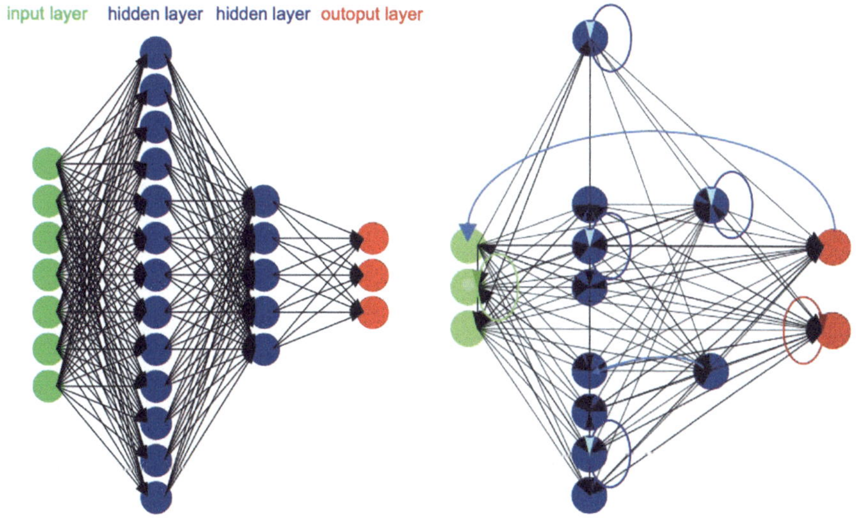

plus local computations.

The output of the single neuron is rather standardized. Apart from biology mimicking ones such as those proposed by spiking neuron models,[11] the commonly adopted input output relation is as in the left picture of the next figure.

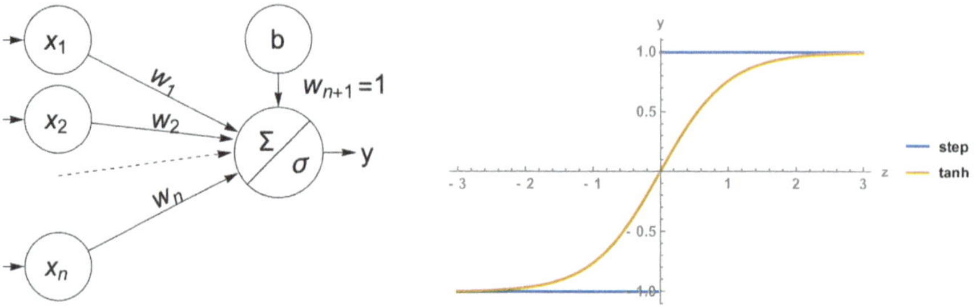

Namely, the signal x_i received by the inbound connected neurons is multiplied by a coefficient w_i, the connection weight, before being added to the contribution of the other inputs and to a constant b, the bias.[12] This *linear combination* (identified by Σ in the picture) produces a value z that is put in input to the activation function σ, whose output y is forwarded to the outbound connected neurons. We are used to denoting both the linear combination as *net* and its result y as the *state* of the neuron. Common shapes of *sigma* are either a step function or a smoothing of it, as shown in the right part of the figure.[13] Many variants are considered in the literature, all sharing elementarity as a main feature. In particular, the circuit on the left part of the above picture constitutes an elementary neural network *per se*, named *perceptron*, connecting input neurons on the left to the unique processing neuron on the right. In spite of its simplicity, it has represented a milestone in the study of brain mechanisms, exhibiting essential functionalities, for instance connecting eye retina to the brain.[14] The oriented graph architecture shown in the second to last picture may be viewed as a sequence of layers where each node is a perceptron; this is why it is normally denoted as *multilayer perceptron*. In the non-oriented graph in the right part, neurons are still perceptrons; as further feature, some edges may start from one neuron of a given layer and may end in a neuron of either a preceding layer or of the same layer, as wellas they may end in the same neuron they start from.

The building block is a perceptron.

[11] Or statistical mechanics components such those embedded in Botlzmann machines.

[12] Obviously, the symbols may vary, like the one of the bias in a previous picture.

[13] Actually the step functions foreshadow computations based on 0/1 bits, which in turn may emulate the spikes produced by the neurons in the brain. Most models of artificial neurons prefer the smooth function because it's more manageable in exploiting the sensitivity of the output to the *net* parameters.

[14] Frank Rosenblatt: Principles of Neurodynamics. Perceptrons and the Theory of Brain Mechanisms.

An essential neural network toolbox

Recalling the above description and related pictures:

- A neural network is a function f of:

1. an external input x with components $\{x_1, \ldots, x_N\}$ collating the inputs to the input neurons,
2. a set of parameters (w, b), with analogous components distributed on the neurons of the networks,
3. a connection layout (the architecture) A,
4. a set of activation functions σ,
5. an activation mode ρ,
6. an external output y with components $\{y_1, \ldots, y_R\}$ collating the outputs to the output neurons,

The first two items are considered the two conjugate variable parts of the function: for fixed (w, b), we reckon the output of f on input x; for fixed set of inputs$\{x\}$, we arrange the parameters (w, b) in an optimal way. Layout A, activation functions $\{\sigma\}$ and activation mode ρ derive from higher order operational processes, like the DNA evolution in living organisms.

- The connection layout establishes inbound and outbound links between nodes of a graph in the role of neurons.
- The activation function is the elementary computations of each neuron, so that a neuron updates its state via the function:

$$y = \sigma\left(\sum_{i=1}^{n} x_i w_i + b\right), \qquad (EPF)$$

let's call it the Elementary Processing Function (EPF), where each x_i either is the output y of another neuron linked by the connection with weight w_i or is an external input piped by this connection.

- Typical activation functions and related graphs are recalled in the figure below, where $a = net$
- The activation mode rules the synchronization of the neuron computations along a time t reckoned by discrete instants (say, seconds, minutes or anything else). Typical modes are

1. Sequential: the neurons enter a loop where their states are updated according to a given order. According to it, at a given time t, the ith neuron takes in input the state of $(i-1)$th neuron that has been updated at time $t-1$, the one of the $(i-2)$th neuron that has been updated at time $t-2$, and so on.
2. Parallel: at time t each neuron state is updated on the basis of the current neuron states, which in turn have been updated as a whole at time $t-1$
3. Random: at each time t a neuron is randomly drawn and its state is updated on the basis of the current neuron states.

Linear	HVS	Sigmoid	RBV
$\sigma(a) = \beta a$	$\sigma(a) = If(a < 0,0,1)$	$\sigma(a) = \frac{1}{1+e^{-\beta a}}$	$\sigma(a) = e^{-(\beta a/2.5)^2}$
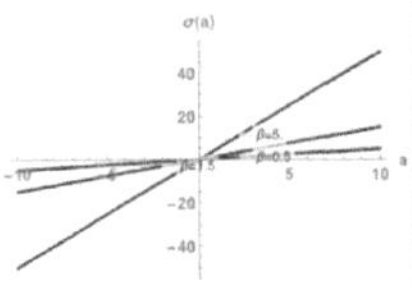	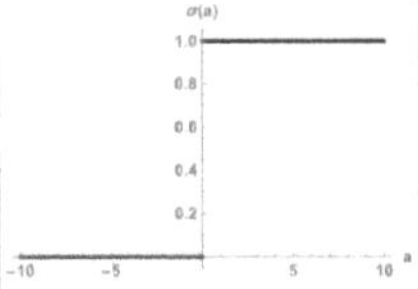	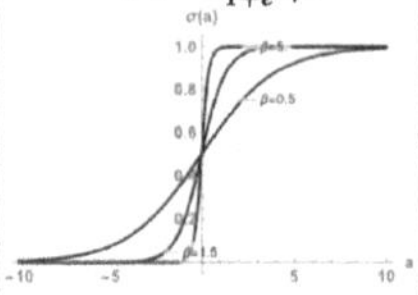	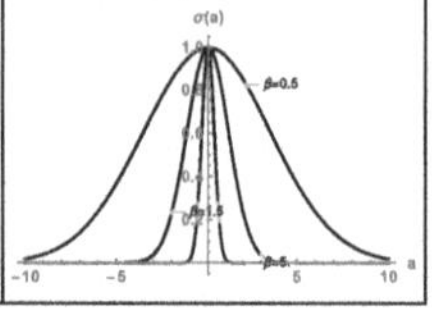

Systems of Ordinary Differential Equations (ODE) recover these dynamics in case of continuously reckoned t.

We may consider a Neural network to be an iterated system. Any subnetwork obtained by dividing the network in two connected parts is a neural network again with the above features.

Summing up, we opened the braincase to give a look to brain and sliced the brain up in order to discover neural networks; then we pretended to mimic brain functionalities by using electronic circuits like the ones in the previous figures, where the signal flows from an input layer to an output layer passing through intermediate ones, the hidden layers, and nodes compute the elementary functions as above. Now the point is: what do we imagine the brain does? *Per se* this is a ill posed question, because we are asking our brain to define itself, which is a typical instance of undecidable problems like the *liar paradox*.[15] Let's list some abilities we expect from it.

The brain should at least:

- solve problems

- *To be a problem solver.* Hence we may see it as an ultra powerful computer that in input any question admitting a computable result, provides the latter in output. Well, the Cybenko theorem[16] states that a neural network like the one in the figure below (a two-layer perceptron) may approximate any computable function provided it is well sized.

[15] Decide the truth value of the following: "this sentence is false". Hence if it is true it is false, if it is false then it is true.

[16] G. Cybenko, Approximation by superpositions of a sigmoidal function. Math. Control Signals Syst. **2**(4), 303–314 (1989).

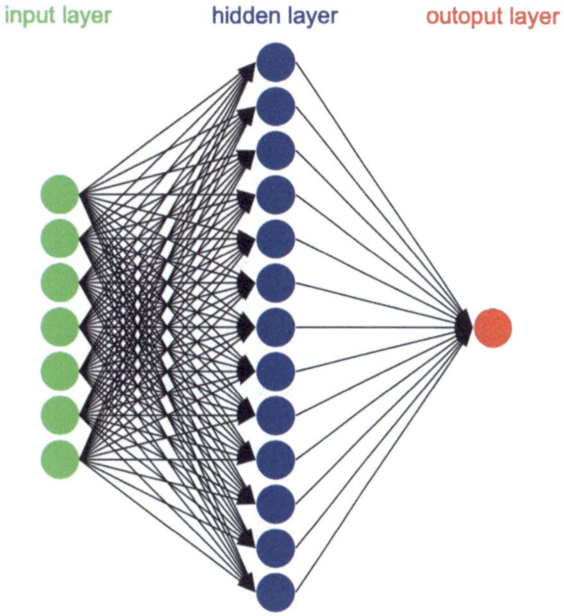

Namely, of the three layers, the neurons of the first one depend of the number of inputs of our problem, while the latter contains the sole neuron outputting the solution. What makes the difference is the hidden layer, where the number of neurons and the accuracy of the connection weights must be so large as to *host* the complexity of the problem we want to solve.

hence, have computational power,

Not yet, however, have we the wizard wand. In fact, *determining the number of hidden nodes and assigning the connections proper weights is a problem that proves in general harder than the original problem we aim to solve.*

- *To gain experience.* I cannot say how much salt and why so much I throw with my *pince* in the boiling water when cooking pasta, but generally it is salted well. That's because I have done this operation so many times, each one, in the early trials, improving the pasta taste. I learned by trial and error. A neural network may do the same. Still better, learning by examples is one of the biggest attitudes of this contrivance, that overcomes the mentioned difficulty of determining in advance the weights of the connections in the above perceptron. Simply, it learns them by examples, like a baby who begins to walk. With this attitude we cover the field of the solutions that are not computed by an explicit algorithm but are dictated by experience.[17]

  - learn new skills

  Why should a neural network learn? The mechanism of the basic scheme of learning from a teacher, otherwise denoted as supervised learning,[18] is sketched below: *you punish and correct the errors of your pupil until he learns to answer your questions correctly.*

[17] Don't waste old women and old men!

[18] But there are other schemes that work as well.

src 6

ANN can do it

The streamline of the related algorithm is the following

Algorithm 1.1 Spervised Learning Template

1. **draw** a set of examples of input/output pairs (x, y) of the problem to be solved—the training set *TRS*,
2. **establish** an error function E that penalizes the network each time on a given x (within TRS) the network outputs a $\tilde{y}$ different from y (for instance by the amount $(y - \tilde{y})^2$)
3. **run** the network on inputs belonging to TRS and cumulate the error E
4. **change** the network parameters, weights w and biases b, on the basis of the sensitivity of E to these parameters (for instance of the derivatives of E with respect to the parameters)
5. **go back** to step 3 **until** E is satisfactorily low

Note that learning is not a deterministic process. Like for human beings, there are lots of reasons why a change of the network parameters may prove inefficient or even counter-productive, like pupil misunderstanding and/or reacting wrongly to the corrections, or in the case of improper punishments, distractions by other phenomena, and randomness in general. So, the result of procedures such as the one above is always a question mark. However, the success of this training scheme says that often, after a proper number of training cycles, the neural network significantly improves its ability to provide correct answers. If pairs in TRS have been picked so as to be representative of the environment of the problem at hand, we expect that E will remain satisfactorily low also when the network is fed with never seen x's, provided they belong to the same environment. *This is the deep rationale of any statistic, as we will elaborate upon later on.*

(provided it is correctly trained).

- invent new abilities,

- *To explore unprecedented solutions.* This is achieved by injecting randomness in the inputs and in the weights of the network. It could be most similar to a mystical ecstasy (by drug or some hypothetical extra-terrestrial entity). But randomness is not magic, as we will discuss in a next chapter. This injection is a common expedient for optimizing a solution. Imagine our error function E depends on x like in the picture below and we simply explore it by always going downward.

– While the red rolling ball would stop in the first local minimum (the black one) if driven by the sole gravity force (and no inertia),
– from time to time (randomly) we might decide to overturn the gravity law, so that the ball has the chance to override the crest on the right and reach a better local minimum (the green one) on the right (the root of a wide family of stochastic optimization algorithms).

fostered by some insanity rooted on the randomness.

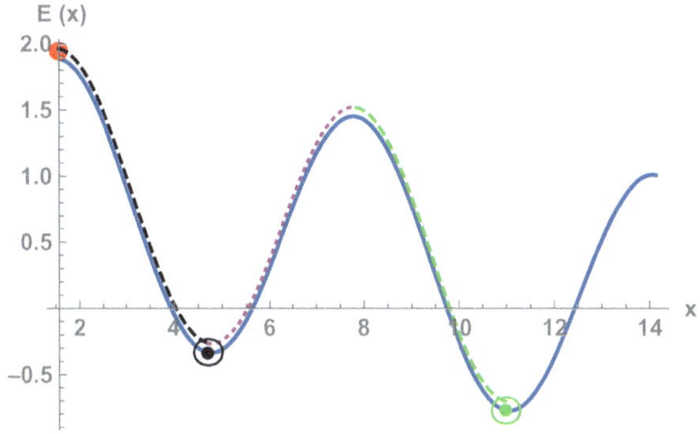

What else? If we are satisfied with these abilities, to which to link all vague attitudes such as intuition, creativity, inspiration and so on, we may opt for a vision of worlds where at the beginning there was chaos, then some neuron-like contrivance began to learn up to the point of achieving the organization of a neural network. Under constant stimuli from the environment, neurons continue to learn, hence to change their parameters (with the mentioned limits on the efficiency of the changes). However, from time to time some neurons sclerotize.[19] This means that they will no longer change the related weights and bias; so they will always produce the same output on the same input, as occurs in the sum of two numbers. We may say that they play the role of a *symbolic function*,[20] where this attitude change represents the essence of the *abstraction* in the mental processes.

The neuron's life.

Obviously, this is a sharp path, that we may name with the evocative slogan *from synapses to rules,* which allows for many variants and caveats. However, it also brings to mind James Clerk Maxwell's vision, according to which "The Universe has bestowed a great gift on man: with its best atoms it has created a part of himself inside his mind to study the rest of himself". We will maintain this as the backbone of our vision in the next pages.

[19] In Medicine, sclerosis means an abnormal hardening of body tissues.

[20] From now on, we call *symbolic functions* those functions (i.e. association of an input within a range to one and only one output within a range) that may be expressed in analytical form through a combination of symbols; by contrast we denote as *subsymbolical functions* those functions for which no analytical expression is currently available to us.

1.4 Further Readings

1.4.1 Biological Aspects

Here my personal lions are roaring loudly. Undoubtedly I've not background in biology, so that discussing genomic questions or biological neurons would be pretentious on my part. Nevertheless, willing to speak of computational aspects, such as evolutionary algorithms and artificial neural networks, requires taking a look at their biological counterparts. Having a guessing about them simply by sniffing on Wikipedia could prove a bit limiting, even though I'm very grateful to the Wikimedia Foundation and its many contributors for the help they provide me in my researches. So, on the one hand I suggest a pair of books that allowed to systematize my scarce knowledge on the matter [7, 9]; on the other one, I highlight a few issues that I have found very inspiring.

1.4.1.1 Genotype, Phenotype, Evolution and Aging

The original meaning of these words refers to the biological sphere. However, we are used to dealing with analogous meanings in computer science

- The *genotype*, as the complete set of genetic material of an individual organism, is the tailoring of that material, denoted as *genome*, to this individual. In turn, genome in humans consists of the mentioned DNA. I always remain harmless when facing the huge variety of DNA components and their functions, just arising from the wording of the four mentioned characters {T, A, G, C}.

  genome <=> coding string

  In Computer Science we are used to giving the three words {genotype, genome, DNA} the same meaning as the code of a given string, hence a program that put in input to a machine outputs that string. Rather, we distinguish two levels of input: the input to the machine, i.e. the program, and the input to the program. The former is at a meta-level with respect to the latter. For instance, the liar sentence "this sentence is false" may be disambiguated if we put the object of the sentence at the first level, and the program which handles the content at the meta-level to decide its truth. So, the sentence "the dog has 6 legs" is false without any doubt.
  In the computational environment both input and output are represented with a binary alphabet. Again we may build an infinite number of input and output using the two characters {0, 1} (actually a still huge but finite number if we limit the number of characters, for instance to 64 or 128), but I don't feel harmless in this case, since I'm not assumed to know (or want to know) their meaning and function in advance.
- The *phenotype* refers to the observable physical properties of an organism. We may consider them the output of the body fabrication machine (allow me this vague term) having the genome as main input, where the output differentiation is related

- phenotype <=> the computation result

to the variety of the genes gathered by genome. Thus, organisms with different genotypes may have different phenotypes. However, organisms with the same genotype may have different phenotypes because of *epigenetic* phenomena. Like the input of our programs, they are additional inputs to the fabrication machine that may modify some properties of the organism, without altering its genotype. It's the favorite kingdom of the short-term memory in the brain.

- *Genome evolution*. While the brain is deputed to face all interactions with Nature during a person's life, genomes of surviving species play a somehow analogous role but in long-term mode, i.e., passing from an individual to his progeny. Hiding details, a species survives if a sufficient number of its individuals is able to cope in a satisfactorily way with their environment. This may be resumed up in a fitness function, one possibly varying over time, which takes values on these individuals, and a probability of reproducing that is positively related to the fitness. Due to the above gene $\leftrightarrow$ phenotype relation, the reproducing individuals will have highly performing genes with high probability. However, the gene shuffle process implemented by meiosis in the sexual reproduction merges the chromosomes of the two parents in such a way that no feasible combination of their genes is excluded as a candidate constituent of the progeny genome; meanwhile only a few (generally one) of these combinations are processed to produce the corresponding individuals. This preserves species diversity, which is always beneficial, albeit within a trend that improves average fitness of the species. Meanwhile, local changes in the chromosome composition[21] gently introduce further variants of the genes and of their layouts in the chromosomes. The alternation optimization and random search steps form a constant duo that we will find here and there in the book in regard both to natural processes and to their artificial emulation.
- The *aging* is a special epigenetic phenomenon. Any reader has a clear notion of aging, even though its main drawbacks are felt in the author's age range (sigh!). An interesting example of intertwining between genotype evolution and aging phenomena may be found in the fight between bacteria and penicillin molecules. The behavior of bacterial cells has always been considered and analyzed at the population level, with all bacteria in a colony acting as homogeneous single entities, totally isolated from the rest of their sibs. Vice versa, on the one hand, biological evidence has recently proven the existence of signaling molecules secreted by individuals in order to convey communication to the rest of the population. On the other hand, biologists no longer consider heretical the concept of bacterial aging at the individual cell level, as well as phenotype heterogeneity and metabolic asymmetry within a bacterial clone, even in the presence of identical genotypes. Note that the single particles have no age in the classical view, while the aging of a population, hence the irreversibility of its life, is decreed by external factors determining a selective death rate. The primary scheme is as follows: bacteria randomly differentiate with time because of random changes in their genetic information, according to the above genome evolution schema; the fittest ones with respect to an

- unavoidable aging

[21] We will discuss them in a grater detail when we will consider the genetic algorithms on Sect. 1.4.2.1.

environment function survive with greater probability at each generation. Hence the growth rate of the fittest individuals strikes the time of the population. Rather, the epigenetic feature of the new scenario is represented by the fact that bacteria may differ in their resistance to an antibiotic as a consequence of their differential ability to produce and secrete an enzyme which destroys the antibiotic. The differentiating process among individuals is connected to an irreversible mechanism that we may associate with their aging. In particular, we may assume that a bacterium has a lever for regulating the partition of its energy between defense and reproductive activity. At each solicitation it is pushed to increase its defensive attitude, unless it is refrained each time its strength is high enough in comparison with a ground state determined by the leverage graduation of the other individuals.

1.4.1.2 The Cerebral Cortex

A brain is much more ordered than expected.

Looking at the enormous number of neurons and their connections in the human brain, we may expect the connection graph to be rather random. Who would manage coherently billions of items? Nevertheless, the evolution of the species generates many ordered regions in evolved brains such as those of humans and other mammals. One of them is the brain cortex. It is a tissue of neurons covering the cerebral hemispheres, which are organized in layers which flow *wired* signals unidirectionally.[22] What comes to mind is an ensemble of multilayer perceptrons like the one depicted on Sect. 1.3. So, while we might associate other areas of the brain to the layout of a rather chaotic social network, this part of the brain is organized hierarchically in six layers, each one having specialized neurons with overall anatomic properties that are different in different areas of the cortex. In the picture below we see three sections obtained with different staining techniques like the one mentioned on Sect. 1.2, so as to enhance the whole cells, their bodies and axons, respectively. The sections are arranged perpendicularly to the cortical surface so that the top layer is the one closest to the cortical surface. The cerebral cortex is considered a main site where information which feeds cognitive functions is processed.

Obviously, the artificial neural network layers have less skills than their biological companions. Nevertheless, the idea that a tight organization is beneficial in the case of some tasks remains highly suggesting.

[22] Actually, besides this kind of transmission along the synaptic wires, which is optimized for precise, high-speed communication, other transmission mechanisms exist in the brain. For instance, volume transmission provides (via diffuse interactions linked to, among the others, oxygen, carbon dioxide, and electromagnetic fields) a communication mode that is temporally slower and anatomically broader in reach, and more suited to modulatory and tuning functions [15]. Moreover, wired transmission is affected by bidirectional signal flow generated by the axons (hence outside the cell body) that deserve some consideration on how to preserve the main unidirectional flow [3].

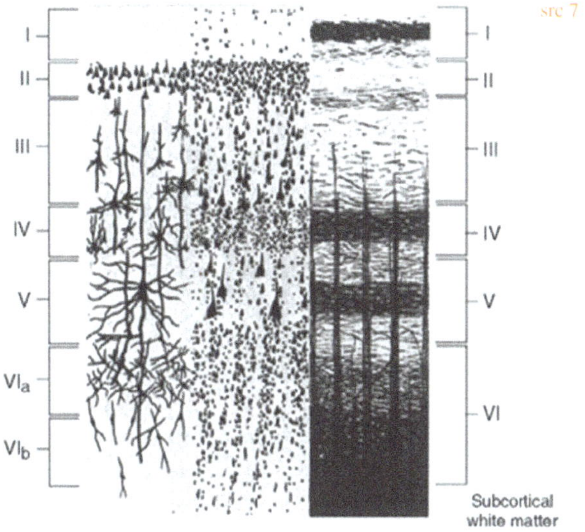

1.4.2 Modeling Aspects

As will be discussed in a greater detail in the next chapters, the last deal in the use of computer systems to support human life is the emulation of activities of biological systems. This emulation is not a novelty, *per se*; we already mentioned the flight of the bird as an emulation target. Related to computational facilities, two main focuses are genome evolution in living organisms, as a tool for prevailing in the environment, and brain activities. They are both templates of activities that are not under our complete control, but that we may exploit for our own benefit. Thus, the former is considered an archetype of the random optimization algorithm, while the latter is the model inspiring a huge family of computational intelligence algorithms. A common goal is to devise artificial systems which perform at least as well as the natural companions.

If Nature can do it ...

A common risk derives from the emulation attitude. Since we are not called upon necessarily to explain our artificial system, but only to describe it and exhibit analogies with the corresponding natural one, our creativity has wide room for maneuver. This is good news in principle; however, the drawback is a proliferation of models and proposals with no guarantee of success, apart from *natural selection*: "Try; if you succeed you survive", that's a key attitude of our times.

let's imitate her.

1.4.2.1 Genetic Algorithms

The block diagram of a template of Genetic Algorithm is the following

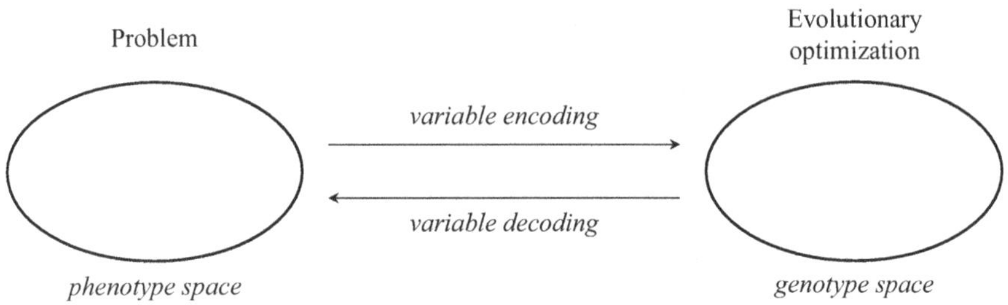

As mentioned earlier, Nature encodes organisms through a quaternary alphabet, computer scientists through a binary one, whereas any coding compliant with *more natural* languages is allowed.

A social improvement passing through:

The crux of the evolutionary optimization process lies in the use of a finite population of n representatives of candidate solutions in a search space $\mathfrak{G}$, whose evolution leads to an optimal solution. For instance, aiming to arrange a team of 11 players of a football team picking from an ensemble of 30, we may start from 10000 binary strings of length 30, where each bit is 1 if the corresponding candidate player is selected, 0 otherwise. Note that the number of all possible strings of length 30 with exactly 11 1s is over fifty million; meanwhile this number keeps growing if we consider the role that can be assumed by the players, hence our population is definitely smaller. We may call those strings genotypes, thus evoking mechanisms which are present in the generation of the human gametes and which will be applied with analogous goals yet possibly different modes.

Key features are *randomness*, *solution* and *fitness*.

chance, and

- Randomness denotes the non complete control of the related operation. Pure randomness is associated to the tossing of a *fair* coin; the tossing may be variously biased in the non-pure release.
- solution is an assignment of values to the coded variables that could solve the target problem.
- Fitness is the result of a function relating the current solution to the cost value of the target problem that we want to minimize; let's call it the *target cost*. The actual expression of this function is up to the user and a challenge for his creativity. Making a parallel with its biological companion, we can say that fitness is linked to the effect of an individual solution, its phenotype, for instance to the performance of a candidate team in our football example above.

meritocracy.

The population-based optimization is an outstanding feature of the evolutionary optimization and is practically present in all its variants we can encounter today. Populations evolve by wise modifications of their elements, that implies a moderate expansion of them, and selection of a subset of them leading back to the original population sizes. This all boils up to a way of linking fitness to the survival probability of an element. Like with other local optimization procedures, we favor elements with greater fitness, even though we do not exclude the possibility that a relatively worse element may be at the origin of a successful modification. The balance between the two criteria is variously pursued by the different methods in the literature and in practice. In a very schematic and abstract way, a computing skeleton of evolutionary optimization can be described as follows:

A survival strategy

1. initialize the genotype population
2. until termination triggers do
   a. generate local variants
   b. evaluate the fitness over the overall population
   c. improve the out-coming population

The engine that drives the algorithm consists of local changes in the coding strings representing the members of a population; for instance, a *mutation* changes the characters of the coding alphabet at some positions in the string; a *crossover* contemplates the twisting of one or more segments between pairs of strings; an *inversion* determines the reversing of characters of a segment. These operations are randomly located in the strings. The general philosophy is that these changes may produce new genotypes without destroying all partial solutions. In turn, this changes the fitness of the resulting strings and primes a process aimed at optimizing the overall strength of the population by renewing its members via a selection procedure that favors, but not exclusively selects, the best fitting ones. With respect to the biological genome evolution considered earlier, the role of the fitness is analogous: namely, to bias the the reproduction probability of the more fitting individuals. The random renewal of the population proves to be different,[23] while still preserving its diversity.

where innovation is realized by local changes.

In the rest of the book we will not deal with genetic algorithms expressly; rather we will variously exploit their abovementioned key features. The interested reader may take an overall view on this family of algorithms in [6].

1.4.2.2 Neural Networks

On the previous pages we have seen the mainstream concept of neural networks. Let me add here some issues and details that might somehow enhance the completeness of the framework.

Technicalities for connoisseurs

- **Probabilistic EPF.** A probabilistic[24] companion of the EPF is the following one

[23] Consider that natural selection passes through two nested populations: that of human individuals and that of the gametes inside each single individual.

[24] We assume the reader to be familiar with the elementary concepts and notations of Probability; in any case, in the next chapter we will have the occasion to review many of them.

$$P(Y = 1) = \sigma\left(\sum_{i=1}^{n} x_i w_i + b\right) \qquad (PEPF)$$

where, now, the Elementary Processor flips a random bit Y with a probability of getting 1 given by the above formula.

- **Learning strategies.** While from observing a sequence of heads and tails we may learn the probability of tossing "head" with quality of learning (for instance its mean square error) that can be sated in advance,[25] with a neural networks things are more complicated. In this book we will adopt for them a
  - *Learning by Examples* strategy, where we use a training set TRS as in the Algorithm 1 to train the network and a Test Set (TTS) to check the quality of learning. Simply, we draw another set of examples from the same set of the TRS, and check to what extent the output of the neural network is satisfactory on these new inputs which have been never seen during the training of the network. A typical satisfaction measure is the distance, in any metrics, of the computed output from the desired one.
  - *Supervised Learning* strategy, where TRS and TTS are sets of input-output pairs of the function we want to learn. It means that we know in advance the output of the function on the examples; thus, like teachers, we may directly control on each example the merit of the output of the network. Vice-versa we tackle an unsupervised learning instance when we don't know in advance the output of the function to be learned. It is the typical case of a clustering instance. It occurs, for instance, when we are called to gather into suitable heaps a set of objects that have been put over time in our attic.
  - *Punishment Learning*, where the cost function of the training algorithm is a function of error by which the output of the network is affected with respect to the desired output (the *Error Function*). It is the most commonly used option. An opposite strategy is the *Reward Learning*; in this case the function is based on the degree of compliance between computed and desired output, hence the *Fitness* of the network, by borrowing the term from the Genetic Algorithms. Error Function must be minimized, Fitness has to be maximized. A useful template of this strategy is *Reinforcement Learning* [14]. Another well-known template, though one less suitable for operational implementations, is the *Boltzmann Machine* [1]. This neural network is endowed with probabilistic neurons ruled by the above PEPF.

- **Backpropagation**. It is a favorite approach for implementing Supervised Learning by Punishment from Examples. Details of the basic algorithms are available on many websites and in papers in the literature; so I won't repeat them here. Rather, I will simply recall the general strategy. It refers to a layered network, hence to an oriented connection graph piping data from the input layer to the output layer, with no loops, by sequentially crossing the layers and updating the neuron states accordingly. The strategy consists of a loop made up of the above forward data flow

[25] We prefer to speak of inferring rather than learning in this case.

that computes outputs from inputs, and a backward piping of the error—revealed in the output layer by contrasting computed output with the desired one—from the output layer to the input layer. During the backward flow an error component is attributed to the single layers and their parameters (connection weights and biases) are modified to reduce the value of the component. The way to make these parameter changes is a distinguishing feature of the back-propagation algorithms. On the whole, they accomplish an optimization task whose goal is the minimization of the error occurring at the output layer. While the algorithm has been designed to cross forth and back an oriented graph, we remark that variants are available for training neural networks whose connection graph includes loops and back connections, as well. These networks are commonly denoted as recurrent neural networks.

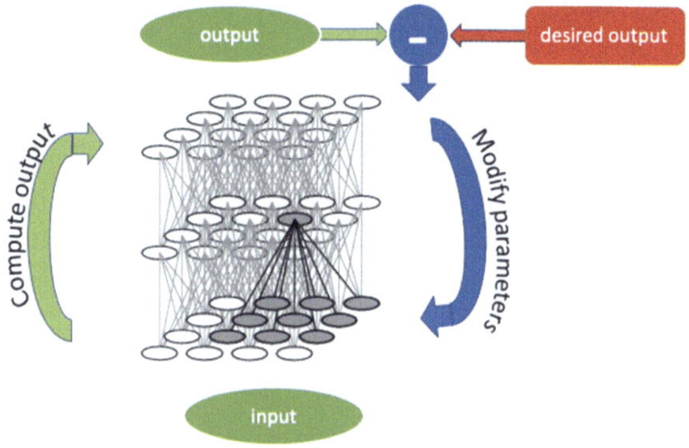

The interested reader may find an overall introduction to Neural networks in [4].

- **Deep Neural Networks.** Deep learning, i.e. learning through a Deep Neural Network, looks to be a must for any computational intelligence research.[26] With this name we denote a unidirectionally layered network, where loops may occur basically around a same neuron (hence its state may concur to form the *net* of the same neuron with an obvious delay). The core of the approach lies in two intertwining goals:

  1. to commit special tasks to the single layers or a group of layers (take, for in stance, a convolution of the input data or an extraction of relevant features), within the overall network goal of learning a function (for instance, speech recognition, as in the vocal interfaces of our personal devices),
  2. to achieve partial and global goals jointly within an auto-differentiation framework.

  This approach has the benefit of addressing the researcher's attention toward the semantics of the operations he is carrying out on the data, meanwhile sparing him from the burden of featuring correct instructions in a proper language to make the computer run correctly in order to perform those operations. So, in the last two decades deep learning has begun to be equipped with standard instantiation and

[26] "An Artist Who Cannot Speak English Is No Artist." is a provoking sentence by the artist Mladen Stilinovic.

training tools. Currently, these tools, structured as a collection of computational objects exploited trough specific interfaces, the APIs, are principally framed on two alternative basic programming libraries: Tensorflow and Pytorch. These APIs have fostered impressive progress in cognitive algorithms and in their application to concrete operational frameworks. Major successes are registered in the image processing, for instance in medicine and in security systems, and in time series analysis, for instance in meteorology and in stock markets. A complete introduction to Deep Learning can be found in [12]. However, all that glitters is not gold. This standardization of the operations, here as well as in many other intellectual tasks, presents some significant drawbacks, which we will discuss in upcoming chapters.

- **Further bibliographical suggestions**
  - A model of aging of a bacterial colony during its fight against penicillin molecules may be found in [2].
  - Models of neuron dynamics along continuous time and EPFs mimicking the ones of biological neurons (such as the Hizhikecvich model) may be found in [5].
  - A book to frame the strategy to overcome crests shown in the last figure of Sect. 1.3 within the family of stochastic optimization algorithms is [8].

References

1. D. Ackley, G.E. Hinton, T. Sejnowski, A learning algorithm for boltzmann machines. Cogn. Sci. **9**, 147–169 (1985)
2. B. Apolloni, S. Bassis, A. Clivio, S. Gaito, D. Malchiodi, An information drawn hypothesis of aging mechanisms in bacterial populations. Nat. Comput. **6**, 33–53 (2007)
3. T. Dugladze, D. Schmitz, M.A. Whittington, I. Vida, T. Gloveli, Segregation of axonal and somatic activity during fast network oscillations. Science **336**(6087), 1458–1461 (2012)
4. J. Feldman, R. Rojas, *Neural Networks: A Systematic Introduction* (Springer, Berlin, 2013)
5. W. Gerstner, W.M. Kistler, *Spiking Neuron Models: Single Neurons, Populations, Plasticity* (Cambridge University Press, Cambridge, 2002)
6. D.E. Goldberg, *Genetic Algorithms in Search*, Optimization and Machine Learning, 1st edn. (Addison-Wesley Longman Publishing Co. Inc, USA, 1989)
7. E.R. Kandel, J.D. Koester, S.H. Mack, S.A. Siegelbaum, *Principles of Neural Science* (McGraw Hill, New York, 2021)
8. K. Marti, *Stochastic Optimization Methods: Applications in Engineering and Operations Research* (Springer Publishing Company, Incorporated, 2016)
9. J. Nolte, *The Human Brain: An Introduction to Its Functional Anatomy/John Nolte; Three-Dimentional Brain Reconstructions by John Sundsten*, 6th edn. (Mosby/Elsevier, Philadelphia, 2009)
10. G. Perea, M. Navarrete, A. Araque, Tripartite synapses: astrocytes process and control synaptic information. Trends Neurosci. **32**(8) (2009)
11. E.K. Schafer, D.P. Lehrman, B. Stevens, The "quad-partite" synapse: microglia-synapse interactions in the developing and mature cns. Glia **61**(1), 24–36 (2013)
12. S. Skansi, *Introduction to Deep Learning: From Logical Calculus to Artificial Intelligence*, 1st edn. (Springer Publishing Company, Incorporated, 2018)
13. M. Solms, *The Hidden Spring, A Journey to the Source of Consciousness* (WW Norton & Co, Manhattan, 2021)

14. R.S. Sutton, A.G. Barto, *Reinforcement Learning: An Introduction*, 2nd edn. (The MIT Press, Cambridge, 2018)
15. K.H. Taber, R.A. Hurley, Volume transmission in the brain: beyond the synapse. J. Neuropsychiatry Clin. Neurosci. **26**(1), iv–4 (2014)

Chapter 2
What I Can Understand, What I Can't

Abstract Abandoning any pretence of knowing the Truth, I discuss in this chapter my approach to delineating a world that's compatible with what I personally experienced. Since there's infinitely more to the world than what I have experienced of it, I look for a possible infinite family of worlds that I may singularly characterize with a distinct degree of compatibility. On the following pages this degree takes the features of a probability measure that can be spent in terms of probability of success of future decisions. The starting point is a uniform distribution, as representative of the absence of information, that we may more suitably reshape in view of observed data and hypotheses on their structure. A special *twisting argument* will allow us to transfer the randomness of the data into the randomness of the compatible distribution that may have generated these data. This delineates a path that disrupts the symmetry hypothesis dictated by the lack of knowledge about a certain phenomenon, leading to increasingly detailed models about it at both probabilistic and physical levels. Finally, to tackle operational situations, we must move from the distribution of worlds to a specific one, as a gross evaluation that will support our decisions. The ways to follow the entire path may slightly differ if we pursue a description goal, typical of statistics, or if we pursue an explanation goal, typical of physics. In any case, the involved computations may be straightforwardly supported by the golden kit, genome + synapses, which we discussed in the first chapter.

During my engineering studies I had the good fortune to learn about the thoughts and ideas of the many researchers who throughout the centuries devoted their lives to improving man's knowledge about how the physical world (what else?) works in terms of universal laws and best practices. The former went in the direction of acquiring *the truth*, the latter represented a backup way of organizing knowledge in local environments when the previous approach looks unfeasible. Then computers came on the scene and things changed. Let's postpone considerations on the last phase. The everyday professional practice pose me the question: "what do I remember of my university studies?", rather: "What is my inner rationale I use of them?". A possible answer is that I remember what's still exciting the network of my neurons because it is compatible with what I observe in my surroundings (the signals I perceive), thus being susceptible of proving useful.

B. Apolloni, *A Few Things I Know About Her*, Intelligent Systems Reference Library 219, https://doi.org/10.1007/978-3-030-94379-0_2

From synapses to rules

According to my own neural paradigm, if some stimulus persists and triggers the same reaction, this denotes subnet sclerosis, hence an abstraction, hence a model. Willing to hit the basic rationale we will start from the root situation where nothing is known by the brain apart a set of signals it perceives. It is the Socrates "docta ingnorantia" thesis: *I know I don't know*. This generate an uniformity feature over the signals that we exploit in the two tasks: describing the signals (statistics), and explain the signals (physics).

In this chapter often we will use concepts and tools belonging to the probability framework. I will not introduce this framework in a systematic way; for that, the reader may reference the considerable literature on the subject. Rather, concepts and tool will come as they need in the discussion. This will have the benefit of not louding the exposition with exceeding formality, but it may prove dispersive time to time. To relieve this drawback, I will provide an essential toolbox of probability in the "further reading" section of the chapter, to which the reader may refer to recap his understanding of the matter.

2.1 A Compatible World

Obviously, I could be an original man who likes considering paradoxes and unfeasible scenarios. But let skip these aspects, which however may not contradict my approach, and focus on the notion of compatibility.

Which world?

Let's start from the notion of parallel universes (multiverses), a framework that has been considered in many disciplines, like cosmology, physics, philosophy, theology, and in fiction too, as a key for solving the indeterminacy of various phenomena. In essence, it refers to variants of our environment which may have aspects and events in common with our own, through similarity or even contrast, and which might represent possible alternatives to our own experience.

- the Miri world?

An example from science fiction is the planet of Miri[1] encountered by the Enterprise crew, a planet that's physically identical to Earth. The histories of the two planets were apparently identical until the 20th century, when scientists had accidentally created a deadly virus that killed all adults but extended the lives of the children.[2]

[1] https://memory-alpha.fandom.com/wiki/Miri_(episode).

[2] Currently, on 2021 this makes me tick something.

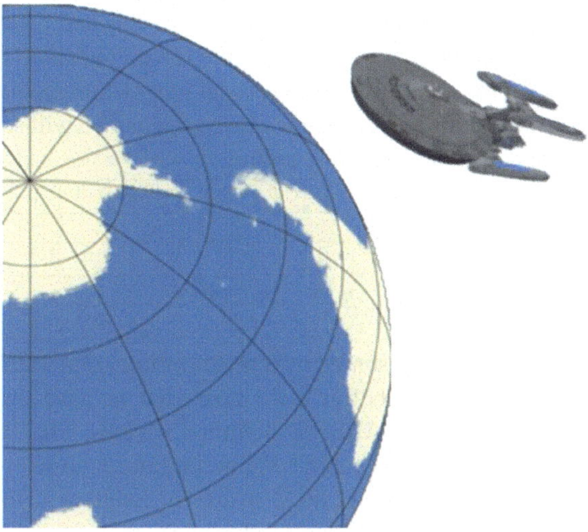

A strong modeling support to parallel universes is provided in physics by quantum mechanics, where nano-scale processes evolve along steps that are superpositions of states which, depending on theory, either really exist or only potentially exist and actually collapse, under particular conditions, into one of them. A typical paradox linked to quantum mechanics is Schrödinger's cat[3] which incurs the superposed states dead and live associated to the quantum evolution of a radioactive material.

- Schrödinger' cat world?

src 8

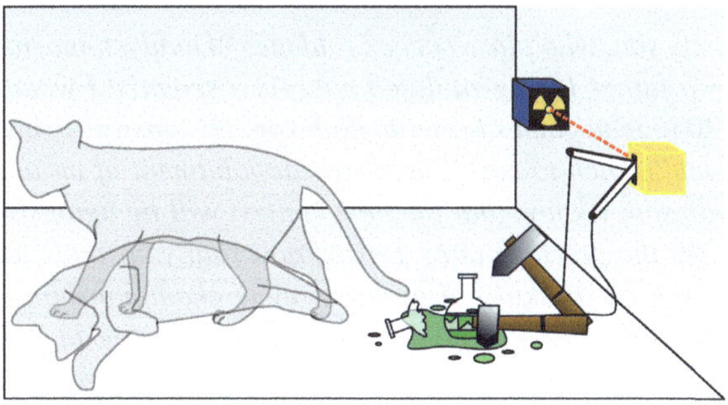

In my experience, multiverses are in the everyday life of any individual who faces operational alternatives. The person's life would be different depending on the occurring options, for instance whether he decides to get married or not, accept a certain job offer or not, and so on. But the individual, since he's always the same person, he would bring his own unique features to any of the different worlds deriving from the above alternatives. Bypassing any deep philosophical discussion, we may sum up the concept with the classic exclamation among university professors: "Ah, if only I had taken that job instead of going into academia!".

- my "if" world?

(I love the roses I missed.)

Let's get down to basics. Here, hoping the reader can follow along, we need to become a bit technical. We start from the most elementary scenario where we want to provide an answer on the success or not of a given venture, for instance our football

[3] https://en.wikipedia.org/wiki/Schrödinger%27s_cat.

team's next game,[4] simply on the basis of its history of wins and losses. We imagine a very poor scenario, where we have no further clues about the result (indeed, we are no football fan). So we must simply organize a set of 0/1 bits including both what we have seen in the past (0 for a loss, 1 for a win) and what we will see in the future. Let us schematize this as in the following picture.

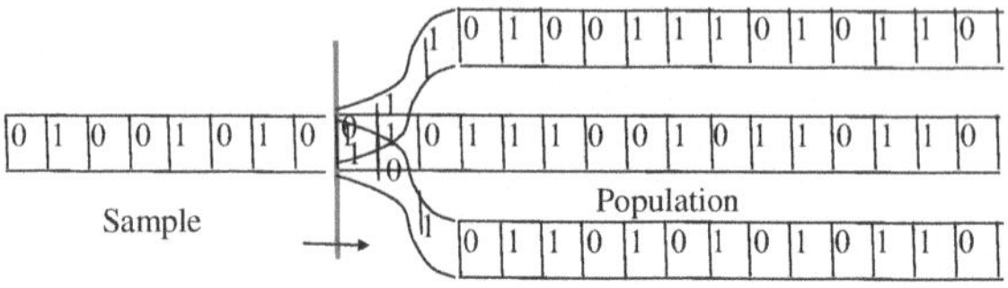

I saw a sample, I guess a population.

Our life re this venture is represented by a huge (infinite in abstract terms) set of strings split in two by a delimiter; before it we have what we have already observed, the *sample*, hence a unique string. After the delimiter we have what we will observe in the future, the *population*; but we don't know the future, so we put all possible histories we might see. This accounts for 2 strings of length 1 if the history ends with the next game, 4 if we will consider two further games, 2^n if we will consider further n games (while we represented a very short subset of them in the picture, for simplicity's sake).

What will be the actual history in the future is a matter of diviner, however we would have something to say about it. During my lectures in Mathematical Statistics I was used introducing this matter with the following tale.

> *Last Saturday my son, who plays soccer, told me: "Daddy, tomorrow we will win for sure." As a father I congratulated him. As a scientist I wondered how many of the last* 20 *league games his team had won. His answer to my question about it was "None". Then to ease him the disappointment of an unfavorable fate, I said: "Dear son, I know that the next contest will be against a team of standard ability like the previous ones. I also know that your team has neither increased nor decreased its skill during this league championship. Now, how can you expect the ability to win a game, which never arose in the previous competitions, to just materialize in the next one?" Of course my son claimed that such an event cannot be declared impossible.*

A win after 20 losses,

Actually, the unique win after 20 losses is not impossible, otherwise we should decree the football league championship over for the year after the 20th game. All the same, I was searching for a result that would be sufficiently *compatible* with the team's past history.

like the history of a bumbling shooter.

Let's set up the following conceptual construct to issue a compatibility measure. The starting point is a linear target of length 1 in any metering scale[5] to be shot by a

[4] Actually, in this book we will often use football as a straightforward working example. However, let me say football is not one of the few things I know about Her. I have no any interest in this particular sport.

[5] Maybe, 1 m, 1 perch, or any other unitary measure.

totally bumbling shooter, so that any point in the target has the same chance of being hit, a situation that we may characterize in the figure below with the fact that the number of targeting bullets per centimeter is almost the same after a long shooting session. Red bullets on the blue bar, the target, reports the results of the first 17 shots hitting the bar; orange bullets over the blue bar refer to the entire session, where the shot results have been aligned into equally sized intervals (20 intervals of size 1/20, forget the upper bullet cutoffs operated by the graphic rendering).

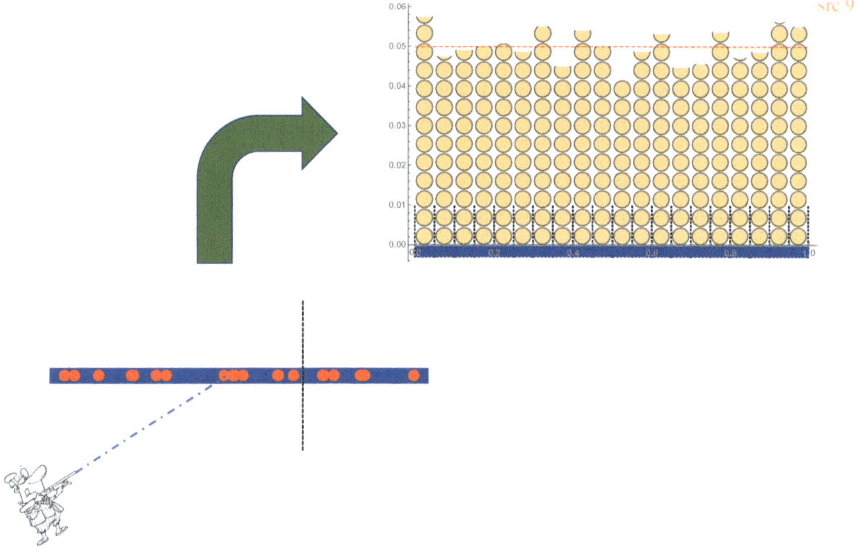

This may be modeled with a *discrete* random variable U assuming values with equal asymptotic frequency, let's call it *probability*,[6] in any one of the 20 intervals partitioning the segment [0, 1]. As for modeling, we may say that there is no any reason for the variable to assume one value rather than another in this segment. *If we divide the probability by the length of the related interval*, we obtain the probability density $f_U(u)$ that results to be constant with u and equal to 1 in the entire segment [0, 1]—a condition that denotes a *continuous U* following a *Uniform distribution law*.

No preferences → equal probability.

Remark 2.1 Now that a variable may assume different values in the same context like in multiverses, we must specialize our notation. We denote a variable to be random if it may assume values independently of our willing and represent it with a capital letter, such as U or X. To complete its definition, we say that U (and any random variable in general) identifies with two sets of values, the former is the set of values $\{u\}$ the variable may assume (the set of its *realizations*, also denoted as *specifications*), the latter is the set of probabilities $\{p\}$ with which those values are assumed. Probability p(u) of a value u is an abstract concept that we approximate

From deterministic to random variables

[6] This value corresponds to the height of the dotted red line in the figure, which only approximately complies with the heights of the columns of bullets, since the latter are in finite number as overall. See the remark below in the text for the notation.

with the frequency with which u is assumed in a neverending sequence of trials—a condition that is denoted as *asymptotic* frequency. More in general we denote *event* e any specification of a function of a random variable, for instance $e = g(u)$. It inherits the probability of the set of all U specifications u's such that $g(u) = e$. For instance, the probability of the event: getting a number less than 5 in the toss of a honest die (all faces have the same probability equal to $1/6$) equals $4/6$. For more formal definitions see the toolbox on Sect. 2.4.1.

Capital $X \rightarrow$ random Small $x \rightarrow$ deterministic.

Remark 2.2 In the picture above the red dashed line has height 0.05 (as the asymptotic counterpart of the observed frequencies) because it refers to the constant probability related to intervals of width $1/20$. It would be 0.01 for a segment of width $1/100$; in general the probability of a segment of width t[7] is exactly equal to t.

We may say that bullets on the target are spread according to a *uniform random variable*, which is modeled as either a continuous variable or a discrete one, depending on the context. Moreover, a further feature of those bullets is that we cannot link the position of the next shots to the ones of the previous ones. This denotes the bullet positions to be *stochastically independent* of one another. Overall, we may say that the shots constitute a *sample* of the random variable U, where sample denotes a set of stochastically independent realizations of U.

No history links $\rightarrow$ independence.

As general notation, a sample S of a random variable X is a set of random variables $\{X_1, \ldots, X_m\}$, where each one follows the same distribution law of X and is statistically independent from the others. We denote with $s = \{x_1, \ldots, x_m\}$ any joint specification of the above variables.

The definition of sample,

The operational counterpart of the independence of two random variables X and Y is that the probability that jointly $X = x$ and $Y = y$ is given by the product of the probabilities of the two separated events, namely: $\mathrm{P}(X = x \;\&\; Y = y) = \mathrm{P}(X = x)\mathrm{P}(Y = y)$. As a consequence, the probability of observing a given sample $\{x_1, \ldots, x_m\}$, let's denote it as the *likelihood* of the sample, equals the product $\mathrm{P}(X_1 = x_1) \times \cdots \times \mathrm{P}(X_m = x_m)$.

and its probability.

2.1.1 The Compatibility Cue

Now we are almost ready to tackle the most elementary question regarding compatibility. We will devote some pages to giving an answer to the question, in a way that is definitely conceptual, but very concrete as well. Consider the vertical delimiter in the last picture and ask ourselves what about the *event* that the next bullet will hit the target to the right of the delimiter. Moreover, we are in so blind situation that we have non any view of the target. We are uniquely interacting with a guy, possibly in the role of adversary, who can position the delimiter anywhere—once and forever—along the blue bar, and the only information passed to us is the number k of bullets

[7] i.e., the probability that U falls in a segment of width t.

that in the past story of m shots did hit the target on the left of the delimiter. The key picture is the following one.

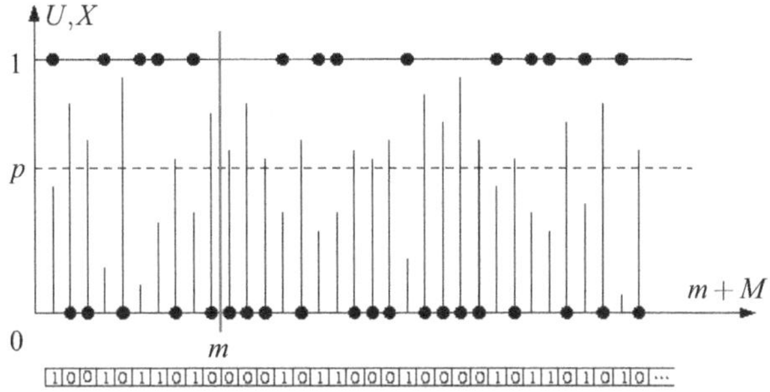

In the horizontal axis we put the sequential index of the shot. The first m shots are those already accomplished; the remaining M are those that we may plan to do in the future (*sample* and *population* in our notation). For each shot we report both a vertical line having height equal to the position of the bullet (its distance from the left extreme of the target) and a boolean variable X equal to 1 if the bullet is on the left of the delimiter, 0 otherwise. The dotted line is at a height p equal to the distance of the delimiter from the left extreme of the target. Let's introduce small formulas to speed up the discussion. From a set $\{u_1, \ldots, u_{m+M}\}$ of bullets positions we get a set $\{x_1, \ldots, x_{m+M}\}$ of the new variable X through the formula (U continuous, X discrete):

From the physics to the probability of an event

$$x_i = \begin{cases} 1 & \text{if } u_i \leq p \\ 0 & \text{if } u_i > p. \end{cases} \qquad (RandomBit)$$

By applying this relation to all possible value assumed by U, we obtain a random variable X assuming only the two values $\{0, 1\}$. In this sense, the u_is are the *seed* of the x_is. What changes with different values of the position of the delimiter p is the frequency with which these values will be assumed in the continuation of the history from the $(m + 1)$th to the $(m + M)$th shot. We don't raise the same question about the sample, since we know the frequency of the values therein. Rather, we may acknowledge that the values of U generating the observed frequency must change with different p's. In principle, denoting by ϕ the frequency of ones in the population, we could also wonder about the order with which the ϕM ones and the $(1 - \phi)M$ zeros follow one another; but it is a nonsensical question because we have established in advance that the u_i's are stochastically independent, hence this order has no any meaning. Let's assume M large enough so that we may speak of the probability p in place of the frequency ϕ, in line with the Remark 2.1 on the previous page, and from the equation (*RandomBit*) we realize that this probability coincides exactly with the height p in the above picture.[8]

Order in the sample is irrelevant.

[8] We recognized in Remark 2.2 that the probability that U falls in a segment of length t is exactly equal to t.

2.1.1.1 The Twisting Argument

The landscape of the compatible worlds which we are investigating is represented in the next picture. The basic strategy is that, if we have no direct information on the position of the delimiter, i.e. height p of the threshold in the above picture, we may take a look at all that can happen.

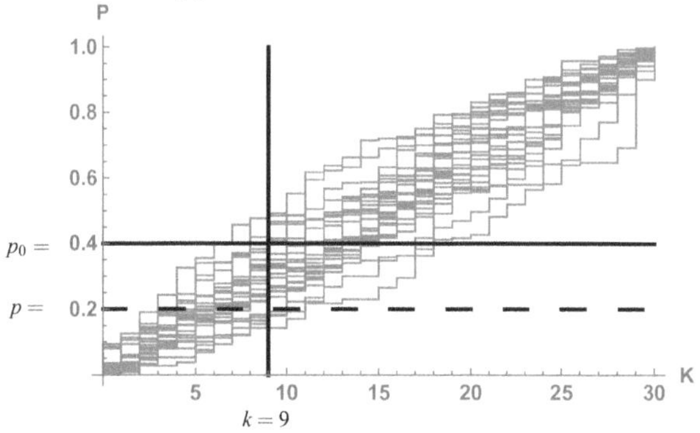

Namely, the stepwise gray lines represent the course of the pairs (K = number of 1s in the sample, P = frequency of 1s in the population) for random stories of shootings with p ranging from 0 to 1. To fix the ideas, we set m = 30 and M = 400. Then, since the shot positions are uniform random variables, but we don't know their values, we draw from U a set of 25 sample uu's of size $m + M$ values, to have an idea of them. Then, since p is unknown, we span the vale of p between 0 and 1, and for each such value we apply (*RandomBit*) to the single uus to get the number $\tilde{k}$ of 1s in the sample and the frequency $\tilde{p}$ of 1s in the population of the shooting story (with usual assimilation of probabilities with asymptotic frequencies). If for any sample we report the course of the pair $(\tilde{k}, \tilde{p})$ we obtain the 25 gray fretlines in the figure below. This is the scenario on which we elaborate three further steps to give meaning to compatibility which overall we denote as the *Twisting Argument* within the framework of *Algorithmic Inference*.

All what may happen.

Logical implication. Let's focus on any one of these stepwise lines, i.e. on any sample of U. The basic feature is the monotonic course of $\tilde{k}$ with $\tilde{p}$, which in turn derives from the monotonic trend of both with p as in (*RandomBit*). Thus, since p is exactly the core of our question, let's assume it to be less than a given value p_0 (we don't know p, but we may establish by us the value of p_0). What we may recognize from the second to the last picture is that if $p < p_0$ then the number k_0 of ones we would obtain when we raise p up to p_0 will be greater or equal to the observed k. Vice versa, if we reckon a $k_0 > k$ this denotes a $p < p_0$. For short

Same effect of the threshold both on the past and the future,

$$(k_0 > k) \Rightarrow (p < p_0) \Rightarrow (k_0 \geq k) \qquad \text{(Logic)}$$

Note that this implication chain is pivoted on two values: the number k of ones actually observed in the first m shots—the sample—and the hypothetical value p_0 we consider as an over bound to the unknown p. The reason behind these implications (that enables us as diviners) lies in the fact that the threshold line in the last figure is horizontal, thus denoting the same phenomena and same effects in the past and in the future. The asymmetry in the implications derives from the fact that:

- raising the threshold parameter in (*RandomBit*) cannot decrease the number of 1 in the observed sample,[9] but
- we can recognize that such a raising occurred only if we really see a number of 1 in the sample greater than k.
  This asymmetry is the cause of the stepwise course of the $(\tilde{k}, \tilde{p})$ lines.

Event inclusion. Now consider the stepwise lines as a whole, as a sample of all possible shooting histories which materializes as *trajectories* of $\tilde{p}$ with $\tilde{k}$. To extend the above implication to each line, let's draw a vertical line where $K = k$ (with $k = 9$ in the picture) and a horizontal line where $P = p_0$ (with $p_0 = 0.4$ in our example). The vertical line intercepts the trajectories at $K = k$, so the ones which cross the line at a $P < p_0$ are those that necessarily are intercepted by the horizontal line $P = p_0$ at a $K \geq k_0$. Referring to all possible stories, the trajectories intercepted by the line $K = k$ with $P < p_0$ represent the event $P_k < p_0$. Here p in the (*Logic*) translates into P because we are referring to random variable, while in the following we hide the subscript k of P as it is tacit in the whole reasoning. Analogously the trajectories intercepted by $P = p_0$ with $K \geq k_0$ represent the event $K_0 \geq k$. The implication of the latter by the former establishes the inclusion relation $(P < p_0) \subseteq (K_0 \geq k)$. We may also state the vice versa but with a different inequality sign as recognized for (*Logic*), so that the entire chain reads

for whatever shot story,

hence, linking next events to past events.

$$(K_0 > k) \subseteq (P < p_0) \subseteq (K_0 \geq k) \qquad \text{(Events)}$$

It says that, apart from bordering stories, statements on K and P are different representations of a same event on the trajectories.

Probability inequality. The random role in the entire discussion is played by the bumbling shooter who is so clumsy that any point of the target has the same probability of being hit. This randomness affects the shooting story which reverses into the randomness of the stepwise trajectories; in turn the latter are exactly what commutes the randomness of the abscissa into the randomness of the ordinate. Thus, the last relation chain we may establish is on the probabilities. Namely, denoting by $\mathbb{P}$ the probability reckoning the probabilities of the shooting histories, actually a meta-probability, we have:

$$\mathrm{P}(K_0 > k) \leq \mathbb{P}(P < p_0) \leq \mathrm{P}(K_0 \geq k) \qquad \text{(Metaprob)}$$

Even with different representations: same events → same probabilities the art of diviners.

[9] Hence, depending on the values of the u_is this number may increase, but it may also remain constant.

The above probability relations derive from the fact that same histories underly both the event $(K_0 > k)$ and the event $(P < p_0)$, where the inequalities derive from the fact that the probability of an event is greater than the probability of any other event enclosed on it, where in turn inclusion is a consequence of the discreteness of K contrasted with the continuity of P.

Probability of probability

Recalling the notion of probability of an event as the asymptotic frequency of its occurrence, we note that in the last inequalities we have three kinds of probabilities:

p	the probability that a bullet hits the target on the right of the delimiter,	corresponding to the unknown position of the delimiter
P	the random value of this position,	accounting for the random shooting histories generating the observed 1's
$\mathbb{P}$	the probability of events on P,	reckoning the overall probability of the ensemble of random shooting histories generating events of our interest according to the observed 1's

Hence $\mathbb{P}$ is a probability of events concerning probabilities. This is why we have labeled the relative inequality as *Metaprob*ability. *Per se* $\mathbb{P}$ is exactly the probability of an event on a random variable. Hence, from now on we will normally adopt the regular symbol P in its place.

We may measure compatibility exactly in terms of the last probability, i.e. by accounting the asymptotic frequencies of the shooting histories that may underlie the two events considered on P and K,

Fixed k, variab. P versus fixed p, variab. K

- the former anchored to what we have really seen, and limited to what will happen in the next observations (the probability my son's team will win),
- the latter anchored to what will happen of the next observation, and limited to what we have already seen.

The two ensembles of histories coincide, apart from some bordering items mentioned before. Note that:

- $\mathrm{P}(P < p_0)$ may be read as the value of a function F_P on the input value p_0. Namely,

$$F_P(p_0) = \mathrm{P}(P \leq p_0)$$

  denoted as the Cumulative Distribution Function (CDF) of P, where for continuous variables $\mathrm{P}(P < p_0) = \mathrm{P}(P \leq p_0)$
- the derivative f_P of this function is the density function[10]

$$f_P(p_0) = \lim_{dp \to 0} \frac{F_P(p_0 + dp) - F_P(p_0)}{dp}$$

[10] A more extended introduction of the density function is found on Sect. 2.3.1.1.

which represents a grade of compatibility of the individual value p_0.[11]

- The probabilities in the two extremes of the Probability Inequality (*Metaprob*) are easy to compute. From them we may derive the probability in the central term.

2.1.2 Compatible Futures: The Cat Remains in Superposition

While at the end of the day, after opening the box we will observe the cat sketched on Sect. 2.1 **either** alive **or** dead, after the observation of a sample the possible prosecutions of the histories remain superposed. What may change are the compatibility measures P of the alternative prosecutions, as a way of containing the *uncertainty* on the future.

Superposition versus compatibility

In our two lead examples, the shooter and the football player, these alternatives are infinite if referred to never ending sequences of shoots and games. We identify them with the probabilities of hitting the target or winning a game, respectively. Thus, the sentences "tomorrow I will win after 20 defeats" and "tomorrow I will win after 19 defeats and 1 victory", map into the two graphs in the figure below, where the left bound of the Probability Inequality (Metaprob) gives rise to the following shapes of the probability $\mathbb{P}$ (i.e. the distribution law $\mathscr{D}$ of P) in terms of $f_P(p)$.

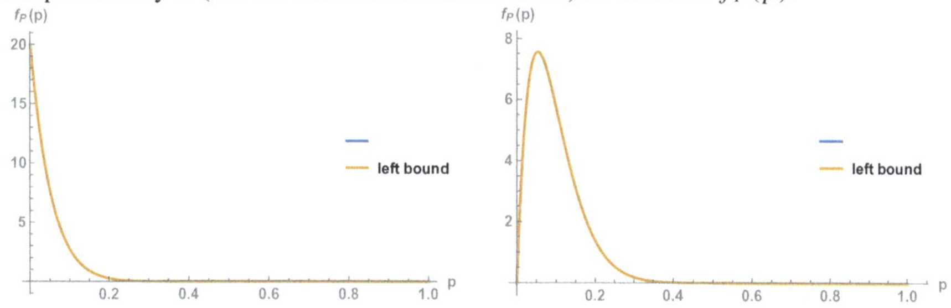

The graphs show that it is unfeasible to determine a priori the fate of my son's next game. But they also establish that a sharp probability of success makes no sense. Rather, we may consider which success probability in the future world is much compatible with the past history, the observed sample, which one is less compatible, where the compatibility measure is provided by $\mathbb{P}$ either in terms of measure of a given

[11] The cultivated reader may distinguish a case such as ours, where P is a continuous variable fully described by F_P or f_P, from the case where P is discrete, for instance after some constraints, so that only F_P maintains its role, while f_P must be redefined as

$$f_P(p) = \mathrm{P}(P = p).$$

From now on the replacement of the above definition is assumed by default wherever necessary. In turn, in case of discrete random variable X, the notation $f_X(x)$ is often simplified in $p(x)$.

event on P, or of a density function on the P values.[12] Actually, in case of 0 wins the graph is definitely drawn to values close to 0. Rather, in the case of one win out of 20 games, the graph is *grossomodo* centered around the value 1/20 that would represent a first common-sense answer to the gambler's question: "What is the probability of the football team winning?" Hence, in these cases tangible compatibility grades are concentrated around 0 and 1/20, respectively. Consequently, betting on these values may make a sense. Would the question follow a history of 7 wins and 13 losses the f_P curve becomes as in the next picture, thus widening the set of much compatible values. For instance, with p in the range (0.3, 0.44), $f_P(p)$ is always over 80% of its maximum.

Quite probably the probability of a win after 20 losses is very low.

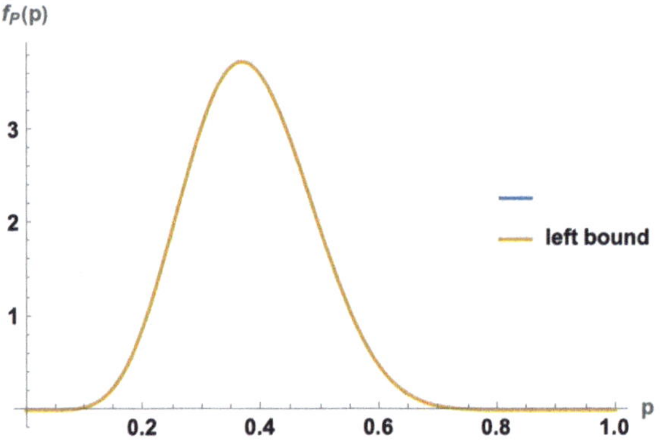

2.1.3 *Indeterminacy and Confidence Intervals*

The future is extremely hard to know, thus opens the door to quite a variety of charlatans, from astrologists to psychics, diviners, gurus and so on.

The fact is, as with the law of energy conservation, there's a conservation law of information, likewise a complementary conservation of unknowledge. If some information is missing, we may exploit our knowledge so as to suitably manage this lack, but we cannot transform unknowledge into knowledge, like with a philosopher's

If you don't know, you cannot invent.

[12] We recall that the probability $P(a < p \leq b)$ is equal to the area under thew the f_P curve delimited

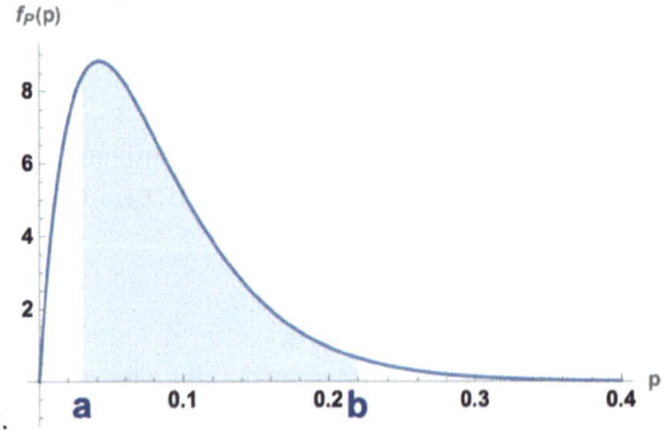

by a and b.

stone.[13] This leads the answer to the gambler question to expand into two intervals, respectively concerning *indeterminacy* and *confidence*.

2.1.3.1 Indeterminacy

The inequality (*Metaprob*) shows that the measure $\mathrm{P}(P < p_0)$ is bordered by two limits. The difference between the two borders is neither enormous nor negligible, as shown in the following picture with the notation $F_P(p_0) = \mathrm{P}(P < p_0) = \mathrm{P}(P \leq p_0)$ since P is a continuous random variable,[14] and introduces a new item in our discussion: *indeterminacy*.

A lack of information may generate an indistinguishable set of distributions.

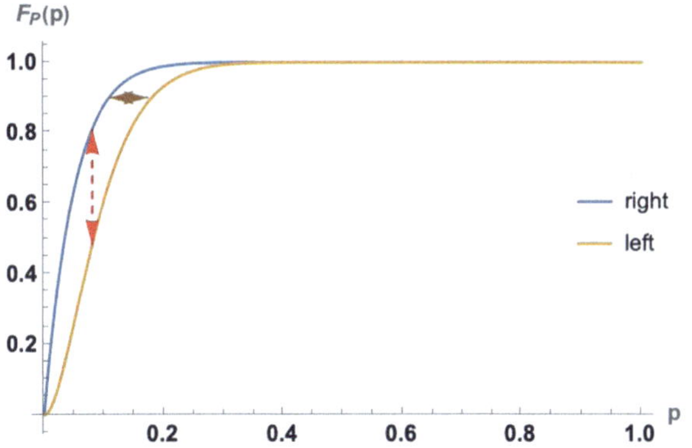

We may appreciate it **either** along the $F_P(p)$ axis, hence by detecting a probability interval, in place of a single value, to be associated to the event $(P \leq p)$ for a given p (red arrow), **or** along the p axis, hence by detecting an interval of values of p to be associated to the equation $\mathrm{P}(P \leq p) = y$ for a given y (brown arrow). Borrowing from mathematical notation with some abuse, we may say that the two uncertainty measures do not commute.

Thus, even if we decide to jump to one of the compatible worlds as an explanation of what we have seen, some thing remains indeterminate. This is not disruptive, but simply a consequence of the granulation with which we decide to observe the world. In our case, we decided to neglect all details concerning the observed football games, apart from the fact that they resulted in a win or a loss.

Remark 2.3 Note that indeterminacy is a common ingredient of our lives. For instance, coming to the shooter stories represented in the shooter picture on Sect. 2.1,

- The bullet positions have equal frequencies in each of the equal sub-segments s into which we partitioned a segment (a, b)—the linear target of the shooter.

---

[13] This is the mythical philosopher's stone, an alchemical substance capable of turning metals into gold (Britannica, The Editors of Encyclopedia. "Philosopher's stone". Encyclopedia Britannica, 10 Mar. 2020, https://www.britannica.com/topic/philosophers-stone).

[14] The graph refers to the bet "tomorrow I will win after 19 losses and 1 victory", since the case 0 win in the past produces a degenerate right bound distribution fixed on 0.

– Vice versa, once we have chosen a granularity s with which to identify the bullet position, for instance in centimeters to take into account the width of the crater generated by the bullet, and attributed the same frequency to each centimeter partitioning the target, we have no way to subsequently locate the bullet in any further partition of the granulation interval s and it makes no sense to forecast with which frequency it may fall into.[15]

2.1.3.2 Confidence

I'm more confident that tonight I'll sleep than that tomorrow I'll become a king.

Besides the above indeterminacy, we may gauge the uncertainty on p in various ways. The most common one is linked to the notion of *confidence*. If my son agrees to meet me at 4 p.m. tomorrow, I'm resigned to wait for him then from ten past four till five o' clock.[16] Vice versa, if the date is with my wife, I expect to meet her tomorrow between ten to four and fifteen past four. This denotes a different confidence I have about the punctuality of my relatives.

In more definite terms, we are used making the notion of confidence precise by introducing the *confidence interval* $I_{1-\delta}$ as an interval (θ_l, θ_r) in the parameter's space where the probability of finding the random parameter Θ inside equals the confidence value, which we conventionally mention as $1 - \delta$.

Coming back to the question of the fate of my son next football match, we must compute the confidence interval in a way that's compliant both with the above definition and with the indeterminacy introduced by the roughness of the binary variable through which we observe the gamers. We start from:

- the inequality (*Metaprob*) to gauge the probability distribution of P between two extremes
- the graphs on Sect. 2.1.2 to appreciate the shape of one of these extremes through a density function
- the graph on Sect. 2.1.3 to compare the two extremes in terms of cumulative distribution functions
- the basic equation $\mathrm{P}(a < X \leq b) = F_X(b) - F_X(a)$ to relate the extremes of the confidence interval to the value of the desired confidence.

and use the whole in a conservative way. So, wanting a confidence of $1 - \delta = 0.90$ we look for extremes (p_l, p_r) such that $F_P(p_l) = 0.05$ and $F_P(p_r) = 0.95$ (see the basic equation above). To this aim, with respect to the left graph in the next figure, I take the intersection of the blue and orange curves (the binding CDFs) with the dashed lines and heights 0.05 and 0.95, respectively. However, I find two intersections

[15] This corresponds to the impossibility of locating p inside the brown interval and $F_P(p)$ in the red interval in the above graph. The association of the bullet position to a continuous variable is an instance of the *continuity commodity* that we will discuss later on.

[16] Here a note is appropriate. "But why do you constantly highlight your son's defects?" a friend asked to me. None of that! I've three wonderful sons; the defects I mentioned are simply nice habits they have (I don't say which ones) that we joke about in our family.

on each line. The conservative strategy is to take the lowest intersection of the two curves with the horizontal line with height 0.05 and the highest intersection with the line with height 0.095. In fact:

- On the one hand a value of p so that $F_P(p) = 0.05$ cannot be less than 0.00256 and, conversely, $F_P(0.00256) \leq 0.05$. Analogously, $F_P(0.216) \geq 0.95$
- Resuming, $\mathrm{P}(P \in (0.00256, 0.216)) \geq 0.95 - 0.05 = 0.9$

In this way, we can gauge the winning probability in the confidence interval interval $I_{1-0.1} = (0.00256, 0.216)$ for $k = 1$ and $(0.177, 0.558)$ for $k = 7$, and use a similar strategy for another confidence interval for a parameter. In conclusion, we cannot establish the exact interval (p_l, p_r) where P falls with probability $1 - 0.10$, because of the above indeterminacy on the P distribution law. Rather, we may assert that $(0.00256, 0.216)$ is a minimal interval that we may certify to have a probability $\geq 1 - 0.10$ of including P. By the way, it's not *the* minimal one since we may find an infinite number of interval with same property simply by shifting left or right p_l and computing p_r accordingly. However, starting from $(0.00256, 0.216)$ we cannot find an interval included in it with the same property. This is a typical local optimization instance that we meet often in our operational life.

Just to find an interval for the random var. *P* and a wise management of the indeterminacy.

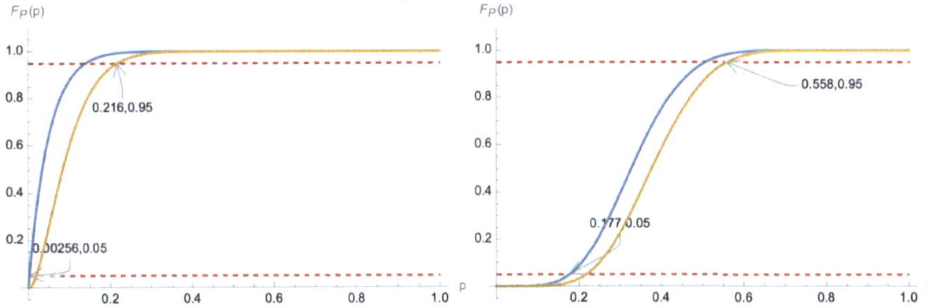

The idea underlying this notion of confidence is that we cannot be sure of gathering a compatible p in (p_l, p_r)—I'm not sure that I will wait for my son for less than 50 min tomorrow—but in the long history of compatible p's I may meet, I will be faulty less than 10% of the occurrences. Note that the probability of success of our inference is not linked to a special family of instances, i.e. "if we solve the same inference problem infinitely many times," but to a given inference strategy, "if we adopt the same inference strategy infinitely many times." Of course the same probability of success persists if we adopt the same inference strategy to solve the same inference problem infinitely many times.

Confidence is a probability to be tossed on long term histories.

2.2 My Own Rudimentary Physics

In the previous section we were dealing with a very poor scenario with no real information for the observer, apart from an ensemble of bits without any structure between them. It is similar to the context of a "Y maze" experiment like in the following figure, where a mouse iteratively put in the main prong learns to keep the left

In principle, the same compatibility problem,

horn where normally it finds food. A feasible task also for an artificial neural networks (and our brain, afterwards) which adapts its weights in response to the positive reward received when it opts for the left horn. Moreover, it's a richer framework than the observation of the cat sketched on Sect. 2.1.

src 10

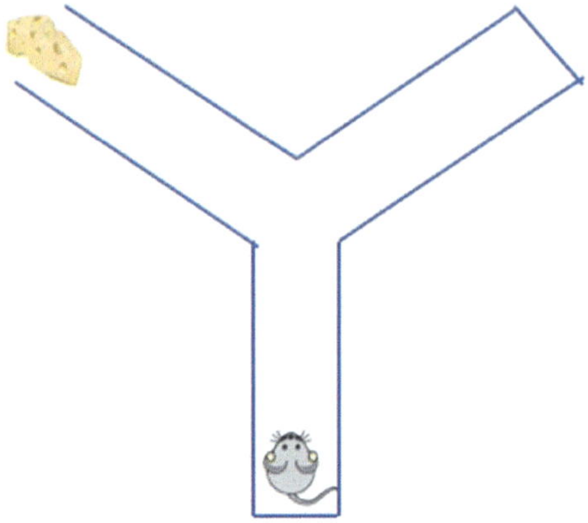

2.2.1 *Cause Effect Dependencies*

but tackled by a smart observer. At the root, the same linear operator *net*,

Luckily, the world is not made up solely of clumsy shooters (shooters anyway), so that our brain may exploit greater information from what we observe. Also in this case, we may start from uniformity as a root of knowledge.

A child learns very quickly that the longer he leaves the tap open the higher the water level in the glass. As an early approach he may reckon a uniform effect on each tick of the clock (second, millisecond, whatever), so that the overall result is the sum of the effects of the individual clock. It is the linear model, the simplest way of cumulating physical events, the same we use to compute the mean value over a sample, the distance covered by our car in a trip lasting three hours, etc. For sure, the model is not proper if we accelerate or brake during the trip; or if the child turns up or down the tap during the filling of the glass. Actually, it proves approximate in the lead example because the glass is not cylindrical (for it widens slightly from bottom up). But it is a very basic relationship that our brain learns and sclerotizes, so that at the end the brain achieves a *function*, i.e. an input output relation which on the same input produces the same output. This may be represented as

$$y(x) = a \times x + b$$

This is the root of almost all functions we use in practice, where if we iteratively replace the atomic effect with the result itself of the function we obtain the quadratic function, cubic and so on. It is instantiated on one-dimensional variables, with plenty of complications when we move to multi-dimensional ones. Henceforth, I will refer to it as my *own rudimentary physics*. Namely

$$
\begin{aligned}
y(x) &= \left(\sum_{i=1}^{x} a\right) + b = ax + b \\
&\qquad (= ax \quad \text{for} \quad b = 0) \\
z(x) &= \left(\sum_{i=1}^{x} y(i)\right) + c = c + \frac{ax}{2} + bx + \frac{ax^2}{2} \\
&\qquad \left(= \frac{ax}{2} + \frac{ax^2}{2} \quad \text{for} \quad b = 0, c = 0\right) \\
w(x) &= \left(\sum_{i=1}^{x} z(i)\right) + d = d + \frac{ax}{3} + \frac{bx}{2} + cx + \frac{ax^2}{2} + \frac{bx^2}{2} + \frac{ax^3}{6} \\
&\qquad \left(= \frac{ax}{3} + \frac{ax^2}{2} + \frac{ax^3}{6} \quad \text{for} \quad b = 0, c = 0, d = 0\right)
\end{aligned}
$$

2.2.2 The Symmetry Classes

assumed to be invariant over equivalence classes.

Obviously, things go differently if the child observes many times the glass exactly 8 seconds after he opens the tap. He falls into a symmetry set whose features do not change, in spite of the different dates at which the observations have taken place. A paradigmatic template of this family of sets is the circle. As a figure, it is invariant with respect to the operation of rotation of any a angle; as a set of points of its border, each element finds in the same set its symmetric with respect to the center. If we move along it with a constant speed we cross a same point at a constant interval of time. All these properties denote a symmetry of the class of circles with respect to various operations, such as rotation, flipping, etc. Said in other words, with the features we put into the play, we have no way of distinguishing between an item and another inside a symmetry set. It is the same kind of indeterminacy we mentioned on Sect. 2.1.3.1 in regard to the discrete random variables: no way to distinguish, no way to infer.

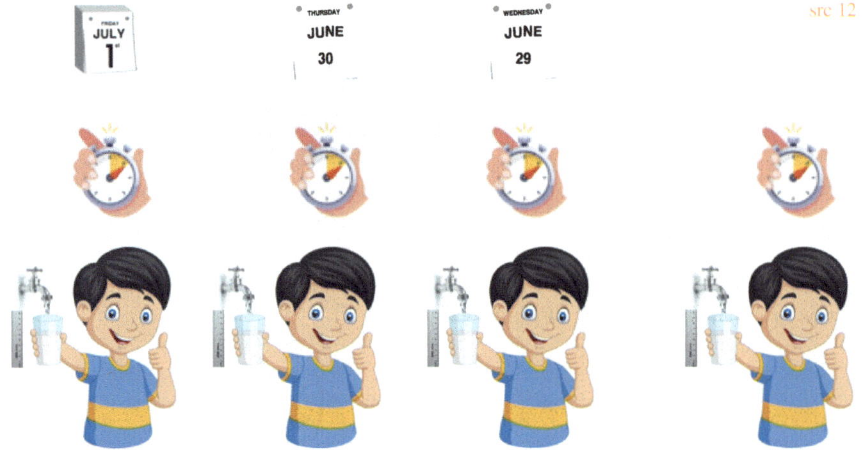

Periodic equivalence classes populate our lives.

Let's remark on the role of the features we exploit. Imagine we are moving at a constant speed along the border of the circle. Hiding the time, for a given x we have no way to distinguish whether we have reached the point (x, y) or $(x, -y)$ in the left picture below. However, if we unfold our loops around the circle, as in the right picture by taking into account the time t, we solve the ambiguity up still preserving the bulk of the symmetry. For instance, red, brown and orange curves distinguish the y-coordinate paths of the trips starting from a same point and lasting a time proportional to $\pi/6$, $\pi/6 + 2\pi$ and $\pi/6 + 4\pi$, respectively. In this way we introduce another big template of signals that are symmetric under translation (with a period 2π), the sinusoidal function at the basis of any periodical signal.

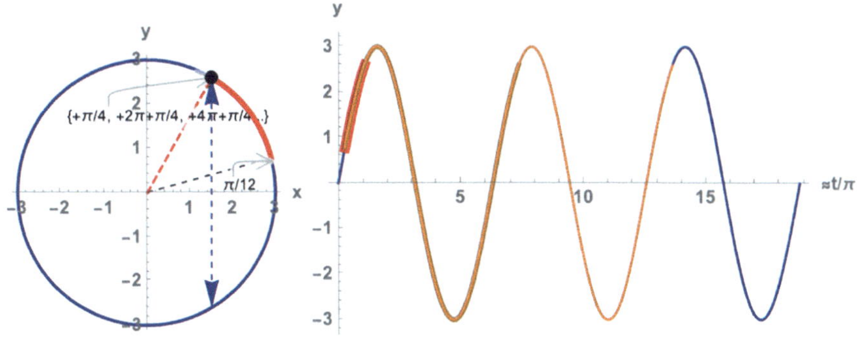

2.2.3 *Much More*

This rudimentary element of physic would represent just the immediate reaction of our brain to external stimuli (better still, what I personally perceive of it) to make them compliant with a compatible world. Physics, as a discipline aimed at modeling our universe, is much more. Modeling entails sophisticated abstraction and representation processes, contrasted with rigorous experiments to support them. Take, for instance, optics theory or quantum mechanics and their instantiation of worlds that are endowed with *imaginary* coordinates and yet remain compatible worlds. Nevertheless, the differential calculus assumed to be the universal language of physics (see point 1 of Sect. 2.3.2) reveals the *net* operations like those on Sect. 1.3 to be at the root of all that.

Cave lions!

Thus, escaping lions (and some colleagues as well), in this book I will limit myself to stretching some aspects of these rudimentary operations, mainly under a cognitive perspective ("don't trust know-it-all people"—according to my mom's advice), remaining definitely grateful, at the same time, for all technology the full discipline has put at our service.

In the Further Reading section we recall also the notion of entropy as a special *net* operation that plays the great role of inner clock of any natural phenomenon. Moreover, in Chap. 4 we will see how the second neural network functionality, the communication, constitutes the keystone of modern physics and of modern society as well.

2.3 Tackling Unawareness Pebbles

Let's imagine our neural network (the biological one being replaced by an artificial one, thanks to their equivalence in our framework) facing a set of new items, for instance boxes of food without label. They are samples coming from a region for which we have to prepare a nutrition plan. Until we don't open the boxes, they are completely equivalent, in the sense that for whatever purpose we want to utilize them, they have the same weight in our brain. Now let's follow the two threads: description and explanation.

2.3.1 The Description Thread

Lack of Information is just a starting point.

Like with the bullets of the bumbling shooter and his target, before opening the boxes, we have no way to decide which one will be most relevant with respect to any criterion (which bullets will get closer to the right extreme of the target), hence we give the same probability of being the most relevant to each box, and, analogously, to each bullet, as a counterpart of the equal probability of each point of the target to be hit by a bullet.

This blind situation may be summed up by the fact that we have no metrics to order the boxes before their opening (neither for the bullet before their shooting). It denotes the poorest condition from an information point of view, shown by the flat shape of the probability density function of the continuous random variable X in black in the following picture. As soon as we have information on a relevant metric, we may reshape the density shape as in blue or green curves, respectively denoting preferences of the random variable U to hit the center or to be biased toward low values.

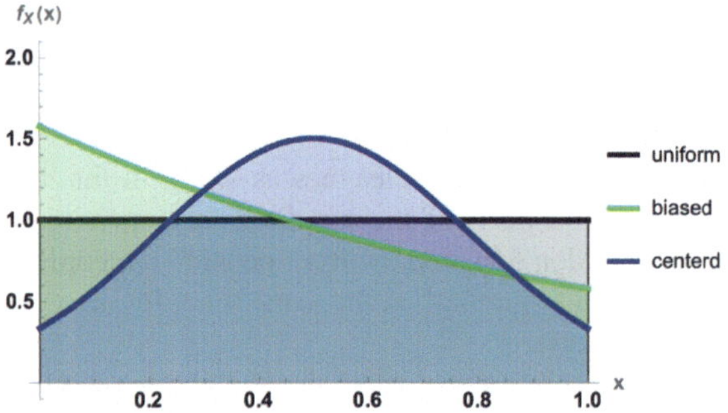

It maybe that:

We may distinguish three kinds of lack of knowledge leading to the discrete uniform distribution $1/n$ (over values $\{1, \ldots, n\}$) and related antidotes.

I've no information,

1. **Lack of information.** Like with the blind boxes, we cannot access any information on their contents. Therefore we list them in an artificial order, just by associating a serial number to each box as the value of a random variable X, and give all of them the same relevance with respect to any metric, probability included which assumes the value $1/n$ over the n boxes. The obvious antidote is to gain information, for instance by opening the boxes.

I'm incapable of exploiting information,

2. **Non exploitability of the information.** Like for the bullets of the bumbling shooter, here we know all about the experiment, but it's not sufficient to distinguish an interval s with respect to another one in the target as for its probability of being hit by the bullets. It's a transfer of unawareness from before to after

the experiment is carried out, raising again the probability $1/n$ over the n partitioning intervals. Here the antidotes must be searched within the experiment conditions.

3. **Induced unawareness**. This is an operational strategy to *sample* data. It may have various motivations.

I want exactly to exploit the lack of information.

- If I won't displease anybody in a lottery, I simply draw randomly the winning numbers, i.e. in such a way that any participant's number has the same probability of being drawn.
- If I don't know the features of a given random variable, for instance the mean income of a population, I draw randomly n specifications of the variable (n citizens), waiting for the wanted features reveal from these values. Again, since I don't know anything (about these features), I give the same relevance $1/n$ to each value. It corresponds to its frequency and, asymptotically, to its uniform probability density.

Pay attention, please. I don't look here for any antidote. On the contrary, I look for suitable procedures to ensure the uniform randomness of the drawn values (see Sect. 2.4.2).

2.3.1.1 Shaping the Distribution Law

From an information viewpoint, the uniform distribution is like an insult, apart from representing a valuable tool in many operational situations. Here we list a series of conceptual issues in the way to identify interesting shapes of the distribution laws. Let's proceed by steps, bearing some formalism along the way. The default option is to consider one-dimensional random variables (scalars).

Toward more informative distribution laws

1. **The infinity commodity**. Consider the shooting history. To get asymptotic stability, the useful quantity to bring to infinity is the number of shots by the clumsy shooter.[17] If we measure his ability profile by the frequencies π of bullets on each of the 20 segments of the blue bar in the shooter figure on Sect. 3.1, the constance of his poor ability is reflected in the fact that those frequencies will not change significantly in a second slot of 10,000 shots with respect those in a first slot of 10,000 shots. This is because in each slot we may assume that all unaccounted factors (hence those beyond the shooter will) determining the shot trajectories had the occasion to be manifested and repeated. Rather, if the frequencies change tangibly we say that the shooting ability changed as well. These conclusions become more acceptable the longer is the slot.
In essence, we rely on fixed frequencies reckoned in unfeasible conditions, i.e. when the slot length is larger than any integer we may mention. We denote them as probabilities and accept that frequencies over, for instance, 10,000 shots are a good approximation of those probabilities, with an approximation degree that

Infinity is like soul: it does not exists but minding about it may prove convenient.

[17] By the way, we are limiting our reckoning to the sole shots that hit the target anywhere.

we may control as an increasing function of the slot length. In the picture below we may appreciate the convergence of the frequency ϕ of 1 with the length $m = Exp[n/3]$ of the shooting history to the asymptotic value $p = 0.05$.

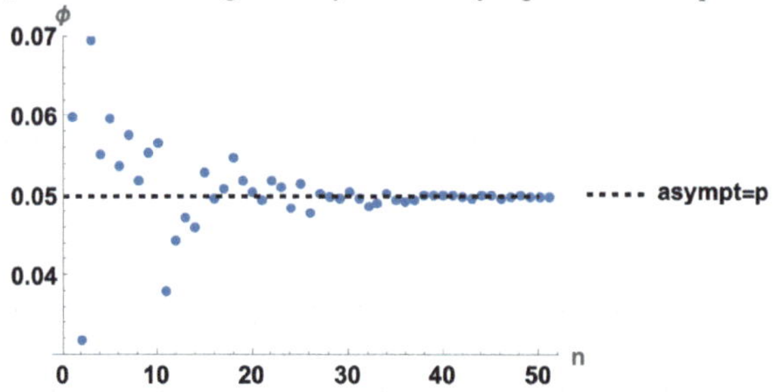

In sharp terms,

- we speak of statistics when we refer to the observed histories and functions of them. If the histories constitute samples, if the sample size is m, the relevance weight of each observation item is in principle equal to its frequency $= 1/m$. (Note that the reference length is m and not the number 20 of partitions of the target segment.)
- we speak of probabilistic modeling when we refer to the infinite slots and their properties.

In conclusion, neither probability nor the infinite slot length may be touched by our hands. But their virtual existence is a *valuable commodity* for our computations.

2. **discrete versus continuous values** Whatever quantity we want to consider, if we want to relate it to a real object, we must measure it (otherwise we stay in the fantasy world). Whatever sophisticated measure instrument we plan to use, we may get only discrete values of the measure. This may appear obvious if we consider the tick marks of a measuring tape or of a kitchen scale. Yet it is confirmed also in the most sophisticated tools involving nanoelectronic phenomena in force of the energy discretization principle at the basis of quantum mechanics. However, the idea of halving a segment infinitely many times, to be more and more precise about the properties of the resulting segments of infinitely small size, offers useful tools for managing functions. We may assume them to be an implementation of the above infinity commodity resulting, in turn, in a *continuity commodity*.

Continuity, as well, is a suitable abstraction.

Thus, continuing on with the bullets matter, we may specialize the above probability measure to infinitely small segments. We may expect that probability, too, becomes infinitely small. Hence, to obtain a tangible value we divide this probability by the length of the related interval, and never stop the halving operation—an asymptotic value again. We denote the result of this operation as the limit of the ratio with the length of the segment going to 0. Namely, let u be the left extreme of the segment, Δu its width, $P_u(\Delta u)$ the related measure, then we define the *Probability Density Function* (PDF) $f_U(u)$ as follows:

$$f_U(u) = \lim_{\Delta u \to 0} \frac{P_u(\Delta u)}{\Delta u}$$

If this limit exists we may say that U is a continuous random variable, otherwise we say it discrete, apart from special cases.
Reversing the above operation leads again to probabilities. Namely, $f_U(u)\Delta u \simeq P_u(\Delta u)$. With analogous limit operations, denoting by $\{X\}_{\Delta u}$ the set $\{^1x, ^2x, ^3x, \ldots, ^*x\} = \{0, \Delta u, 2\Delta u \cdots * \Delta U = u\}$ we may define

$$F_U(u) = \lim_{\Delta u \to 0} \sum_{\{X\}_{\Delta u}} f_U(^i x)\Delta u = \int_0^u f_U(x)dx$$

which takes the probability of hitting the segment $(0, u]$ and in general

$$\mathrm{P}(a < U \leq b) = F_U(b) - F_U(a)$$

As may be implemented for any continuous random variable, with this *continuity approximation*, we may represent the densities of some random variables as in the pictures of the second to last figure and may compute the probabilities of some events simply as the area under the density curve as sketched in Footnote 12 on Sect. 2.1.2. Rather, if we refer exactly to the uniform random variable U we have:

$$f_U(u) = 1 \quad , \quad F_U(u) = u \quad \text{for every } u \in [0, 1] \quad \text{and} \quad \mathrm{P}(a < U \leq b) = b - a$$

3. **Sampling mechanism**. Once the boxes are opened we have features of their contents that we may use to order the boxes through metrics on them, hence on values (of these metrics) to which to connect events of interest to us. The metrics' values x are resumed by variables X (which may assume these values). The simplest event we may consider is a threshold event T_t which, in the case that X is a one-dimensional variable, is the Bernoulli experiment whose result *Res* is described as follows:

If I can mea-sure, then I can move far from the uni-form model.

$$Res_T = \begin{cases} \text{success} & \text{if } x \leq t \\ \text{failure} & \text{if } x > t. \end{cases}$$

Since X is a random variable we transfer the probabilities from X to T in the obvious way of summing in P(success) the probabilities of all different values x_is of X that fall below the threshold.[18] With respect to the *net* operation on Sect. 1.3, it refers to a neuron input 1 on each synapse, weight $p(x_t)$[19] on each

[18] With this last operation we complete the exploitation of the properties of the probability measure, which are formally synthesized by Kolmogorov axioms that will be reported on Sect. 2.4.1.
[19] Recall the notation $p()$ introduced on Sect. 2.1.2.

connection and bias = 0. In particular if we start from the total unawareness condition—namely: uniform probability distribution on a segment [0, 1], hence X equals the uniform variable U seen before—the above summation may convert into an integral, having $f_U(u) = 1$. In this case, $F_U(u) = u$, so as we can define the random variable T, denoted as the *Bernoulli variable*, which assumes values $\{0, 1\}$ (respectively in the case that $Res_T =$ failure and $Res_T =$ success) with these probabilities

$$P(T) = \begin{cases} p & \text{if } T = 1 \\ 1 - p & \text{if } T = 0. \end{cases}$$

where $p = F_U(t) = t$.

Rare and crowded events: the two basic families of probabilistic models, still rooted on the uniform one.

The magic U. It is worth to note that all principal models of random variables derive from the Bernoulli experiment. In particular Poisson and Gaussian variables are the ancestors of two families of variables related to sum of a large number of Bernoulli variables respectively with an extremely small p—hence related to the *rare events* like the burning out of bulbs—and with appreciable p and a rescaling factor—related to the *crowded events* such as those determining the actual position of a balance scale.[20] As a first intuition, we may figure events linked to people in a square. The first row in the following figure represents rare and crowded conditions, respectively. The number of people walking inside the square in the left part may be described through a discrete random variable following Poisson distribution, while the distance between any two neighboring walkers through a continuous random variable following a Negative Exponential distribution law (f_{number} and f_{distance}, respectively in the second lane if the figure). Despite, it makes no sense to consider this distance in the right picture, neither the total number of people (a huge number in the picture that is infinite in the model); rather we may approximate the people density (for instance, number of men per square meter) with a continuous random variable that follows a Gaussian distribution law (f_X in the figure).

[20] Typically, if we put a same object several times on the same balance we will obtain different values each time. Goldsmiths knows this and normally uses three values for a same chunk to decide its weight. The many events concurring to a single measure are linked to factors such as weather temperature and humidity, maintenance of the various scale components, speed with which the object is put on the balance, etc.

src 13

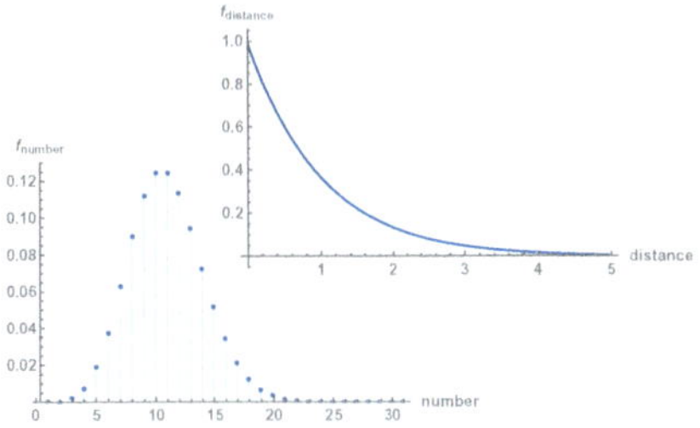

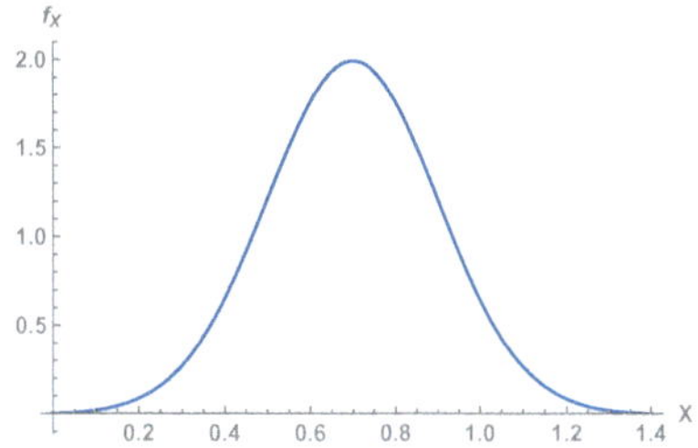

The physics of the unawareness This box is a rather technical one. So, the lay reader may skip it. Let's visualize the rationale with the following bookshelf metaphor illustrated in the following picture.

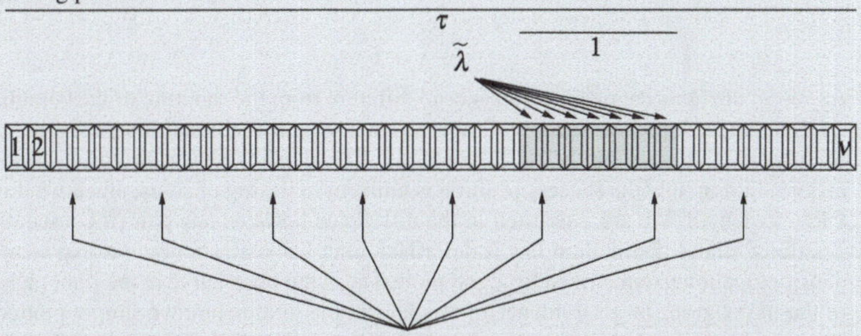

In the following picture we see a shelf of a library of length L containing N books. The librarian is interested on peaking Hemingway books, which are in number of m in the shelf, that we may split in the product $\nu \times t$ where ν is the number of books (all equal as to appearance and thickness) in a linear meter of the shelf and t the length of the shelf where we could confine contiguously the books. Actually, it is the librarian's first day of work, so he cannot guess where a Hemingway book is located, so that his elementary strategy is to pick at random on the shelf (a practice now suggested with some persistence in choosing politics). Note, given the librarian unawareness, his probability of success π is the same if the target books lined up to one another or simply scattered on the shelf. Namely, $\pi = \nu \times t/N = (\nu/N) \times t$. Now, we may expect that ν/N, let's call it *density*, is extremely small, since the shelf length L is very large; thus the probability of success depends on the *size* t of our target. However in the case he fails the first try (picks a book and discovers it is not Hemingway's) he can do further n trials, to be compared with the overall number N of books on the shelf via the ratio n/N, denoted as *intensity*. The interplay of these three parameters:

density, size and intensity, distinguishes the above families of rare and crowded events with respect to two variables:

- K, which reckons the number of successes of the librarian's trys, hence $K = \sum_{i=1}^{n} X_i$ where X_i is the Bernoulli variable revealing the success of the single trial
- V, which measures the distance in number of trials between two consecutive successes.

Given the high number n of involved Bernoulli experiments—henceforth we will call infinite for short each time this attribute can be considered by abuse as a "static" property in place of the limit of a series—it happens that the frequency of 1's in the sequence and their number in a prominent subset of them do not commute: either the frequency vanishes and the number is finite or the frequency is non vanishing but the number is infinite, as we will see in a moment.

Namely:

1. if the density is low and the size is as well, we may rely on an intensity large enough to obtain on average a tangible number $\mu = {}^{\nu\times n}/_N \times t = \lambda \times t$ of picked Hemingway[21] books after his trials. The distribution law of the number K of picked books is the Poisson law with parameter λ that we may interpret as an asymptotic value of ${}^{\nu\times n}/_N$ for moderate ν and a balanced infinite growth of both N and n. The figure below illustrates how the shape of the related density function, now denoted as P_K to highlight the discreteness of its argument, depends on λ).

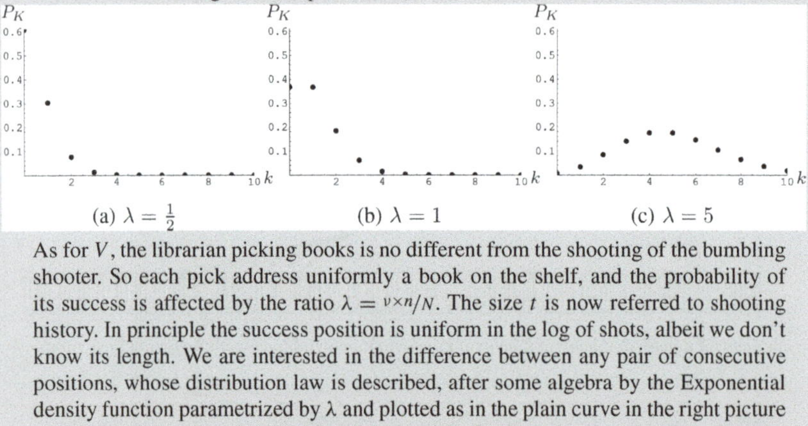

(a) $\lambda = \frac{1}{2}$ (b) $\lambda = 1$ (c) $\lambda = 5$

As for V, the librarian picking books is no different from the shooting of the bumbling shooter. So each pick address uniformly a book on the shelf, and the probability of its success is affected by the ratio $\lambda = {}^{\nu\times n}/_N$. The size t is now referred to shooting history. In principle the success position is uniform in the log of shots, albeit we don't know its length. We are interested in the difference between any pair of consecutive positions, whose distribution law is described, after some algebra by the Exponential density function parametrized by λ and plotted as in the plain curve in the right picture of the next figure. To get evidence of this law, in the first picture we simply plotted a sample of a random variable uniform in [0, 1]; then, we sorted the sample and plotted in the second picture the differences between consecutive values. Points in the right picture are the rescaled frequencies of these differences.

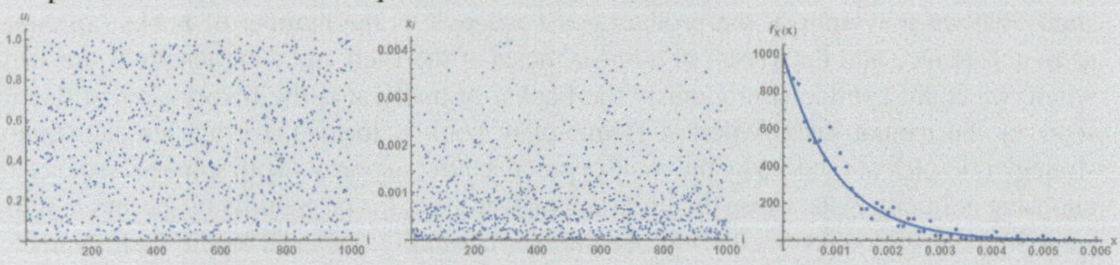

The whole synthesizes, like any other random variable, an unawareness transfer between events. In the case of the Uniform variable we have no idea of the next value we will observe in the interval [0, 1]. In the case of the exponential distribution, referred for instance to the time duration of a lamp, we have no reason to say that it occurs at a specific failure time, with the additional ignorance of a right extreme of this variable (in theory the failure may occur at a time greater than any one we may fix).

2. if the density is large, hence m is comparable with N, then the above product $\lambda \times t$ is huge, namely asymptotically infinite, thus we must rescale it with another large quantity. To recap: the original target variable was $K = \sum_{i=1}^{n} X_i$, then, since K assumes too large values (say, for instance $K = 10^{12}$) then we divide for a *fixed* amount d of the same order. So, whatever the physical quantity$D$ to which d may referto, for instance the surface occupied by the above crowd group of people, the ratio K/d reports a density of successes of the Bernoulli experiment per unit of D. Given the eventually high values of the terms of this ratio, its approximation to a continuous variable is quite natural. Thus we come to the new variable Y whose name is Gaussian, and whose example of plot is in the next figure.

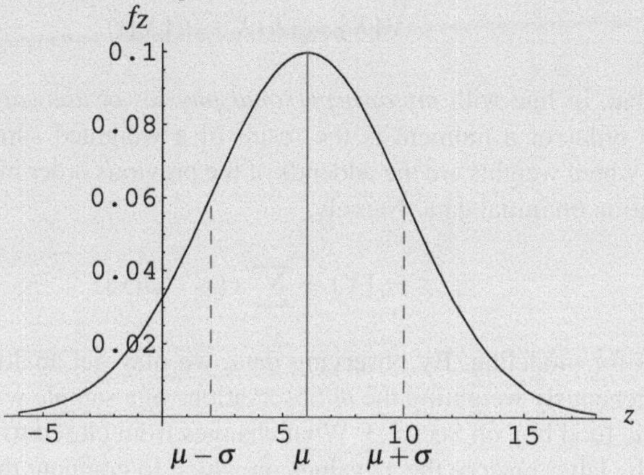

4. **Gross evaluation** Though in a way different from in the quantum mechanics multiverses, when it faces the set of values offered by a random variable X of interest to us, our brain must collapse them in a limited number of values, to save memory space and react promptly to solicitations that will drive our own universe. Since we aim at a single value, the common way is to take an average of the set, i.e. the weighted sum of the set values, where weights are the probabilities of the individual ones. This average, denoted as expected value $E[X]$, is computed as

One sharp number at the end of the day.

$$E[X] = \sum xp(x)$$

for discrete X, where we use the specific notation $p(x)$ of its density function (see footnote 11 of Sect. 2.1.1.1). While its corresponding expression for a continuous X is

$$E[X] = \int x f_X(x) dx$$

[21] To be mathematically picky, he could peak more than once a same book, since he's committed to reinsert in the shelf every picked book.

Hence, we deal again with a *net* operation. More in general we speak of *moment* $m_r[X]$ of order r of a given random variable as

$$m_r[X] = \sum x^r p(x)$$

where $m_0[x] = 1, m_1[x] = E[X]$ and the ensemble of all moments of any order uniquely identify the X distribution law. Special mention goes to the m_2 moment, which we'll meet many times in the book as a dispersion index. In particular when applied to the variable $X - E[x]$ it is called variance V. Namely:

$$V[X] = (E[X - E[X])^2.$$

Note that, in line with *my own personal physics of unawareness*, an increase of one order of a moment is the result of a weighted sum (a *net* operation again) where weights are the addends of the previous order moment, like a deep meditation on initial data. Namely,

$$m_r[X] = \sum x(x^{r-1} p(x))$$

This is for modeling. By observing data, we may get analogous moments by homogeneously weighting the m observations of a sample with weight $1/m$, like with the food box on Sect. 2.3. What changes from closed to open boxes is that from the latter emerge the x_i values on which to compute the moments, albeit we do not fully know their distribution law. Namely

$$M_r[X] = \sum x_i^r \frac{1}{m}$$

We say that M_rs represent estimates of m_r, again a task for neurons, whose assembling may produces estimate collapses of more complex functions of X.

A fourth dimension along which to analyze the data.

5. **Processes** A common way to analyze physical phenomena, including life, is through a sequence of frames, like in a movie. In this sense, we speak of processes. Random processes indeed are described by a set of random variables $\{X_i\}$ whose distribution laws change with time. Rather, we may say that it is precisely this change of the distribution law that evidences a time elapse. Time, *per se*, is not necessarily the one marked by a clock; rather it is exactly the *dimension* along which the frame changes. For instance, the evolution dimension could be latitude (the distance from the equator) if we deal with climate conditions, or the quality of the materials if we deal with bulb duration distribution laws. In essence, we may think of a sphere, like in the picture below, that we may rebuild by superimposing slices that have been cut in any direction.

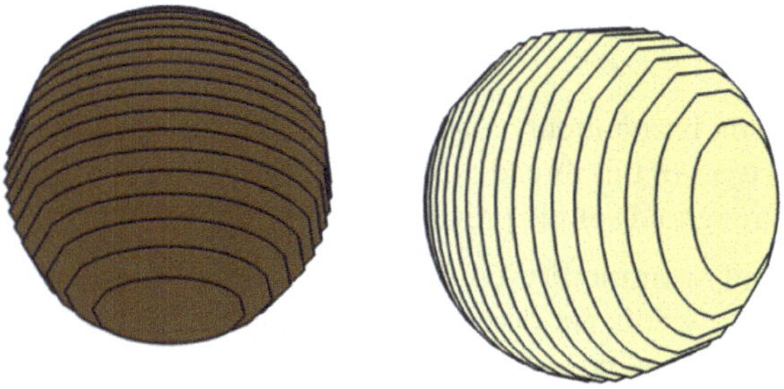

Note that the selected clock acts at a higher abstraction level. Thus, we may speak of T as a random variable, for instance Exponentially distributed as in the previous picture, for instance describing the time duration of a bulb. We have a process when the parameter λ of this distribution[22] changes with the life itself of the bulb.[23] This may occur in practice because the type of lamp in question, not unlike our body, undergoes an aging process.

6. **Environment** In spite of claimed unawareness, stemmed by the uniformity of the root variables, the environment we have discussed up till now possesses a wealth of knowledge, which is represented by the awareness of its borders. This allows us to state that $m_0 = 1$. In other words, we know the set $\mathfrak{X}$ of the values the random variable may assume and claim that $\int_{x\in\mathfrak{X}} f_X(x)dx = 1$ or $\sum_{x\in\mathfrak{X}} p_X(x) = 1$.

   We may speak of probability only if we know the domain to which it refers to.

   In the case of unawareness of the borders, we cannot exploit these global equations and must abandon the probabilistic framework. Rather, we may refer to $f_x(x)$ as unbounded weights, possibly individually constrained to fall in proper intervals, for instance in the interval $[0, 1]$. This is the *fuzzy set* environment that we will exploit in the next chapter. However, this is not the poorest environment. Actually, there are many grades of granulating the data. For instance we may be enabled to establish the x values only by intervals, etc. As in political affairs, at the very worst there is no end.

   Environment → conditional distribution.

   Beyond the extremes of the interval of values a random variable may assume, we deal with its conditioning mainly through the Bayesian approach, where we consider the probability of an event given that an other one we know it occurred. For example, the probability we'll encounter a traffic jam in a particular square, given that today it's raining out (see Sect. 2.4.4).

[22] In greater detail, the density function of T is

$$f_T(t) = \lambda e^{-\lambda t} \quad \text{with} \quad t \geq 0,$$

where λ is the failure rate.

[23] The liar paradox is always lurking.

2.3.2 The Explanation Thread

Observe a phenomenon and ask *why*.

Explaining data is commonly a matter of physics. Actually modeling and statistics are the main tools of this thread as well, albeit with a more bias toward the modeling. Hence the strongholds of the previous thread reappear almost identically as follows

1. **The infinity commodity** Differential calculus is the formal language of physics, whose base words are derivatives and integrals. For instance, the motion of a body is described through the three functions of time t: acceleration a, velocity v and space s, linked by the equations

$$v(t) = \int a(t)dt + v_0$$
$$s(t) = \int v(t)dt + s_0$$

Iterated applications of linear functions, for a start.

For constant $a(t) = a$ they represent the continuous counterpart of the first two reiteration formulas on Sect. 2.2.1. Namely[24]

$$v(t) = at + v_0$$
$$s(t) = \frac{1}{2}at^2 + v_0 t + s_0$$

Give a the role of uniform density in the uniform probability model and we have the same *net* scheme of a neuron with equal unitary values of synapses and weights a. Reiteration formulas (replace the atomic effect with the result) is the typical way of elaborating on new variables, again to put in the same net scheme.

Continuum for philosophers, discrete for implementers.

2. **Discrete versus continuous variables** In spite of some initial familiarization hardness, analytical computational tools for continuous variables are definitely more comfortable than their discrete companions. This is why continuous models are preferred for delving into physical problems. On the other hand, measurements are discrete, as mentioned earlier, and numerical computational tools deal with discrete variables by definition. Hence we are constantly concerned with the approximation introduced by shifting from continuous to discrete frameworks and vice versa.
3. **Gross evaluation** Like with the descriptional variables, explaining variables as well need a synthesis in order to determine one from among multiverses. In classical mechanics this is commonly performed through the *net* operator, giving rise to the moments of various orders where the weights are the explaining variables and the inputs are linear coordinates of the points they refer to be raised to the index of the moment. A big diversity arises from the fact that, while the probabilities are dimensionless variables (numbers), the explaining variables are physical entities possibly spanning multi-dimensional spaces. This makes

At the end of the day, we must take decisions.

[24] Therein the term $ax/2$ derives from the discretization of x, namely, $\sum_{i=1}^{x} i = (x(x+1))/2$.

the *net* operations more complex, specially if referred to higher order moments, and the interaction between variables demanding as well. Hence with reference to the linear stick of length L, if we affect points with a (linear) mass density function $\rho(x)$ which is analogous to the probability density function, the generic moments m_i are computed as

$$m_i = \int_0^L x^i(\rho(x))dx$$

with:

a. $m_0 =$ the overall mass
b. $m_1 =$ the mass center (barycenter) $\times\, m_0$
c. $\tilde{m}_2$ i.e. the second moment related to the coordinates $x - m_0$ is the moment inertia of the system.

If we consider a unidirectional force, like gravity, hence by multiplying the mass density by the gravity acceleration g, we have the configuration of forces like in the picture below. Namely, we considered two configurations of the forces: distributed and concentrated on isolated points, respectively. In the former we considered a biased distribution of ρ, so that the force diagram is the green one in the picture. In the latter configuration we assumed the forces to be equal, somehow reminiscent of the sampling operation over a random variable X, while the points have been drawn from a variable with probability density which is biased like in the green diagram. In this case the barycenter coincides with the sample average on X and the total force is represented by the plain red arrow. Its position is slightly different from the plain black arrow representing the total force cumulating the green diagram, while a comparison on their modules depends on adopted scale factors.

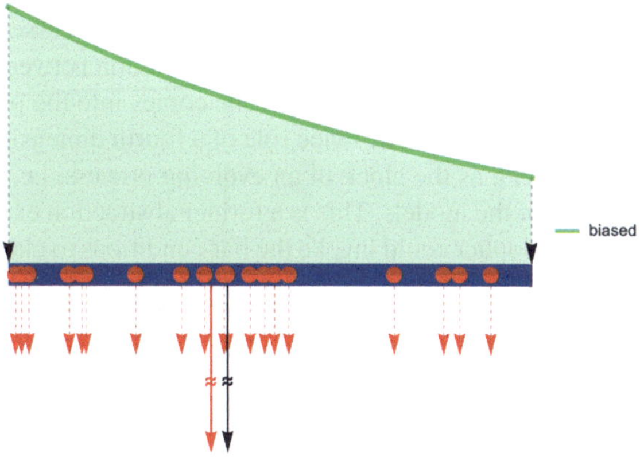

Shifts from this simple instance call for complications that are beyond the scope of this book.

net mechanisms are at the basis of many physical phenomena, too.

4. **Sampling mechanism**. In this case too, we obtain new variables by iteratively managing the action chunks with net computations. Obviously, we are not saying here that this iterative mechanism is capable of explaining the entire world and all its many intriguing questions. Rather, I highlight some of its power.
   - Taking the gravity attraction for granted, all of the classic mechanic is a matter of net operations, whose combination may provide very challenging problems such as the *Three Body Problem*[25]
   - Logical operations such as *and*, *or*, *xor*, or the computation of extreme values may be tackled by a known composition of *net* operation, hence by a neural network (see more details on Sect. 3.4).

5. **Processes** *Per se* a process is the evolution of the state of a phenomenon. As mentioned before, in statistics we identify the state with the probability distribution of the values that some questioned variables may assume. In classical physics we collapse the distributions exactly to the value we observe of those variables in an approximate framework—passing from capital letters (sets of variables and their probability distribution) to small letters (single values from the above sets)—where we may assume exact relations, the state equations, between the values assumed by the variables in the various instances of the process evolution. Thus, the linear relation between the level l of water in the glass of the figure on Sect. 2.2.1 and the filling time t:
$$l = qt$$
   absorb the random effects on L and T, which derive, for instance, from the oscillation of the pressure in the water pipes, the variation of the sections of the glass or the approximation with which we register the time.

   Deterministic processes → state equations,

   In effect, the above equation establishes a regression problem where we look for a nominal tap flow q that renders the equation *essentially* true, for instance on average. In this framework we speak of a physical processes as a phenomenon as a whole, distinguishing its static aspects—relation between variables at fixed time—from dynamic aspects—where time comes into the play. In fact, like for the description, time plays the twice role of a fourth dimension, entering in play and, at a *metalevel*, as the clock of an evolving process, i.e. the clock affecting the parameters of the models. This is a former abstraction exercise of the human brain. The philosopher could invoke the liar conditions to claim that the problem

[25] In classical mechanics, the three-body problem consists of taking the initial positions and velocities (or momenta) of three point masses and solving their subsequent motion according to Newton's laws of motion and Newton's law of universal gravitation. No general closed-form solution exists, as the resulting dynamic system is chaotic for most initial conditions, and numerical methods are generally required.

is ill posed. The implementer could face no small difficulties on synthesizing the behavior of the evolving system through relations hiding the meta-level of the time or some other clocking variable.
Things become harder when we cannot forget the above random effects, thus considering relations of the form

$$L = QT$$

entailing relations between random variables. This is the core of mechanical statistics, if we assume that at each instant a relation is stated between single values of the three variables in question. Rather, it is the core of quantum mechanics if we assume that at each instant all the values of the variables interact with one another, though we are allowed to see a single result at a time of these interactions.

stochastic processes, per usual.

6. **Environment** The notion of environment plays a special role in physics, well represented by the claim of engineers: "They are the boundary conditions which substantiate a problem".
Setting boundary conditions is the way to specialize the solution of a general physics problem to our practical instances. Such conditions may represent true nightmares for the problem solver when they are very specific.
As we will see in the next chapter, specific boundary conditions and inadequacy of the gross solutions that hide random perturbations called for the help of computers toward the end of the last century.

Who is environment free?

2.4 Further Reading

As anticipated in the introduction of this chapter, I have tried to discuss of basic concepts without introducing a highly formal framework. While not a good approach from a scientific perspective, it may prove more accessible by a lay reader. To partially repair the drawback of this strategy, here I add a *toolbox* to sum up the main probabilistic tools we have used, along with some more details about the chapter's subject matter.

2.4.1 The Probability Toolbox

Here I won't provide an exhaustive introduction to probability. I will simply sketch out some strongholds of the subject. There are many books where the interested reader may find a rigorous and complete introduction to probability theory. I just suggest a couple here which, albeit not the latest, provide a robust perspective on the matter [7, 13].

At the basis of the most accepted frameworks of probability theory there is the following definition.

The three Kolmogorov axioms.

Definition 2.1 A *probability space* is a triple $(\Omega, \Sigma, \mathrm{P})$ where Ω is a non empty set (the sample space, i.e. the set of of all possible results of the experiment under question—let's call them outcomes), Σ a sigma-algebra on it[26] and P a *probability measure* represented by a function from Σ to the unitary interval [0, 1] such that:

1. $\mathrm{P}(A) \geq 0$ whenever $A \in \Sigma$ (P *is a measure on the event algebra* Σ...);
2. $\mathrm{P}(\Omega) = 1$ (... *whose unit is the measure of the outcome space* Ω, *i.e. the certain event* ...)
3. $\mathrm{P}(\cup_{i=1}^{n} A_i) = \sum_{i=1}^{n} \mathrm{P}(A_i)$ whenever $A_i \in \Sigma$ and $A_i \cap A_j = \emptyset$ for each $i, j \in \{1, \ldots, n\}$ with $i \neq j$ and n possibly going to infinity (... *and is an extensive measure*).

Probability Space to frame, Random Variable to describe.

A *probabilistic model* is any way of specifying the probabilities of a probability space. An efficient way is represented by the random variable. Broadly speaking A random variable X is a bijective function from a partition of Ω and a subset of the real line $\mathbb{R}$.[27]

---

[26] For an outcome space Ω, a collection $\mathbf{A}$ of events over Ω is an *event algebra* if

1. $\Omega \in \mathbf{A}$ (Ω *too is an event*);
2. $A^c \in \mathbf{A}$ whenever $A \in \mathbf{A}$ (*what is not contained in* A *can be described through elements of* $\mathbf{A}$), where A^c is the *complement* of A with reference to Ω, i.e. the set $\Omega - A$ constituted by the elements of Ω not belonging to A;
3. $A \cup B \in \mathbf{A}$ whenever $A \in \mathbf{A}$ and $B \in \mathbf{A}$ (*compounding however pieces of* Ω *of interest to us still gives an event*).

The event-algebra is a Σ algebra if property (3) is substituted by

(3′) $\cup_{i=1}^{+\infty} A_i \in \mathbf{A}$ whenever $A_i \in \mathbf{A}$ for each $i \in \mathbb{N}$ (compounding an infinite number of events in Ω still gives an event)

From point 2 we derive that, an observation which does not fit with the elementary events listed in Ω belongs to its complement, i.e. the empty set $\emptyset$ w.r.t. Ω. Since $\emptyset = \Omega^c$, this set is an event of any $\mathbf{A}$. Thus, the *anomalous event* is an event of any $\mathbf{A}$ as well.

[27] A less demanding definition refers to any function X from events in Ω, provided that the set $\{A : X(A) \leq r\}$ belongs to Σ for any $r \in \mathbb{R}$.

2.4.1.1 Random Variable Descriptors

Here below we list a few functions with which we are used to describing a one-dimensional random variable and relevant properties of them. For a more complete and rigorous enunciation of this matter, the interested reader may refer to various introductory books. I simply list the following three, which accompanied my education (while a bit dated, the texts are very clear) [7, 10, 13]. For an introduction to the Algorithmic Inference, I also cite two papers of mine [1, 2]. I have tried to keep this book self-contained, at a cost of some approximation and some fuzziness here and there. The hope is that the functions I list here can be managed by the reader with a clear idea of their rationale and no entries to dramatic misunderstandings, i.e. without adding new bricks to the *simplification conspiracy* (see Sect. 4.4.4). In summary, this might only serve to spark the reader's curiosity so he'll go on to delve properly into the topics discussed.

In particular, at the bottom of the next table we sketch the *life* of a function of a random variable. For instance, if X is the result of tossing a honest die, this entails a former mapping from the results of tossing the die to values of x belonging to the set $\{1, 2, 3, 4, 5, 6\}$, and X assumes these values with identical probability $\frac{1}{6}$. Assume $g(x) = (x - 3)^2$. Hence the variable $Y = g(X)$ assumes values in the set $\{0, 1, 4, 9\}$ with probabilities $\{\frac{1}{6}, \frac{2}{6}, \frac{2}{6}, \frac{1}{6}\}$, respectively. For instance, Y assumes value 4 when $x = 1$ and $x = 5$, thus inheriting the probability $\frac{1}{6}$ from both these events on X. Analogously for the $F_Y(y)$, as sketched in the graph therein yet with reference to different X and g (namely: $F_X(x) = 1 - e^{-2x}$, with $x \geq 0$, and $Y = X^2$).

In the table I adopt the symbol $\mathfrak{X}$ to denote the support of the discrete random variable X, i.e. the set of values this variable may take. Then, for simplicity's sake, I will assume these values to be sorted in ascending order. By contrast, the support of X continuous will be assumed to be $(-\infty, +\infty)$ by default (possibly with set of values having probability density null).

Random variable descriptors

mode	definition	properties	graphs
discrete	The cumulative distribution function (CDF) of a random variable X is defined as: $F_X(x) = \mathrm{P}(X \leq x)$	$0 \leq F_X(x) \leq 1$ for $\mathfrak{X} = \{x_1, \ldots, x_n\}$ with $x_i \leq x_{i+1}$ $F_X(x_i) \leq F_X(x_{i+1})$	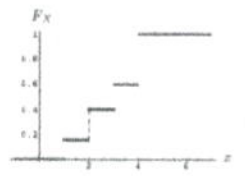
	The probability distribution function (PDF) of a discrete random variable X is defined as: $f_X(x) = \begin{cases} \mathrm{P}(X = x) & \text{if } x \in \mathfrak{X} \\ 0 & \text{otherwise.} \end{cases}$ PDF is also represented as $p(x)$ and also denoted as density function	$f_X(x) \geq 0$ $f_X(x_i) = F_X(x_i) - F_X(x_{i-1})$ $f_X(x) = 0$ for each $x \neq x_i$ for each $x_i \in \mathfrak{X}$ $\mathrm{F}_X(x) = \begin{cases} 0 & \text{if } x < x_1 \\ \sum_{r=1}^{i} f_X(x_r) & \text{if } x_i \leq x < x_{i+1} \\ 1 & \text{otherwise.} \end{cases}$	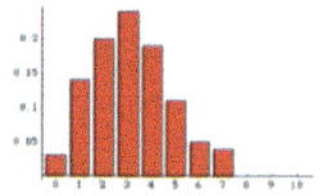
	The r-th moment $m_r(X)$ of a discrete random variable X is $m_r[X] = \sum x^r p(x)$	$m_0(X) = 1$ $m_1(X) = E[x]$ $m_2(X - E[X]) = V[X]$	
continuous	The (CDF) of a random variable X is defined as: $F_X(x) = \mathrm{P}(X \leq x)$ (same as for discrete)	$\lim_{x \to -\infty} F_X(x) = 0$ $\lim_{x \to \infty} F_X(x) = 1$	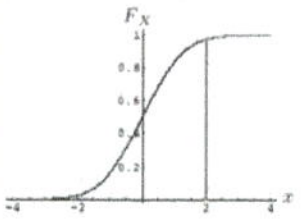
	The density function (PDF) of a continuous random variable X is defined as: $f_X(x) = \frac{dF_X(x)}{dx}$	$\mathrm{f}_X(x) \geq 0$ $F_X(x) = \int_{-\infty}^{x} f_X(t)dt$	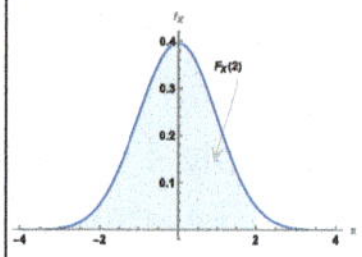
	The r-th moment $m_r(X)$ of a continuous random variable X is $m_r[X] = E[X^r] = \int x^r f_X(x)dx$	$m_0(X) = 1$ $m_1(X) = E[x]$ $m_2(X - E[X]) = V[X]$	
Functions of a random variable			
	Quantile X_α: $X_\alpha = \; x$ such that $F_X(x) = \alpha$	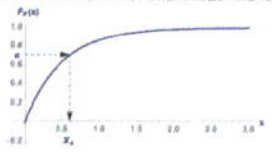	
	$Y = g(X) \Rightarrow$ $f_Y(y) = \sum_{\text{all } x \text{ s.t.} g(x) = y} f_X(x)$	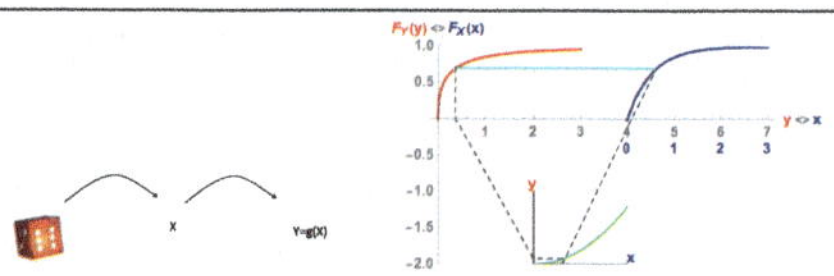	

2.4.2 Random Number Generators

At first chaos came to be.[28] The idea of a primary disorder seems to be familiar to human beings since millenniums. Then the other early idea was that a superior entity (God, Demiurgo, Brahama[29] and so on) was so skill to establish order on the whole stuff, getting the cosmos,[30] and we live happy and grateful to him. Obviously, pursuant to the will of any such entity, anything is possible, both chaos and the current order of nature. So the question "and who generated the chaos" is switched off by default. According to the preferred approach of this book, we may consider chaos as a population of Random Uniform variables.[31] We use this variable because

U as archetypal chaos.

1. There is no reason why we observe one value of this variable in a given interval preferably than another one (of same width).
2. There is no way to shift from this condition if we have observed a long-as-we-want record of previous values of this variable.

Since these conditions denote essentially a lack of knowledge on the phenomenon we are observing through this variable, it is not hard for us to imagine a game situation where a contender passes such kind of values to his antagonist who is unable to forecast, even to a minimal extent, what will be the next value.

By inverting the perspective, a contender might adopt moves based on such a random number to keep his adversary from unfavorably conditioning this decision.This is the same rationale as in flipping a coin on a binary decision totally unclear of us. It's the rationale of a large family of *probabilistic algorithms*, where the adversary is the instance of the problem itself and we want to avoid meeting an instance particularly unfavorable to a non- random strategy we may have decided to adopt. *When in doubt, flip* may prove the operational counterpart of the ancient sentence *In dubiis abstine*[32] (When in doubt, abstain). A basic survey of the probabilistic algorithms may be found in [8].

When in doubt, flip.

Thus we need random numbers in many instances of our lives (computations included), and we cannot continuously flip a coin to get them. Thus any computer operating system includes a routine for recursive generation of a specifications' sequence of uniform random variable U over the [0, 1] with a decided accuracy (number of bits to represent a specification). The rationale is to endow the player with a computer which runs an algorithm that is known to him, but not to the other contender.[33] Rather, aiming to discover the algorithm, the latter should involve a so huge computational effort that this task proves much too costly, when not absolutely unfeasible.

Random as it needs.

---

[28] Hesiod cosmogonia.

[29] Yahweh is the creator demiurge god of Nature, called Brahma in Hindi and Khudah in Arabic.

[30] Exactly the order in ancient Greek.

[31] Here chaos is in the acceptation of the Greek expression chôra as a shapeless space in which material traces of the elements are in disordered motion—nothing to do with chaotic systems.

[32] Latin legal sentence.

[33] Rather, the algorithm is known, except for some key parameters.

A popular example of the algorithm is the congruential algorithm shown below. The output is a sequence of given-approximation numbers, hence binary strings of fixed length, that

Unifor-mity is easy;

1. are cyclically repeated but with a cycle so long that the user will never double any number during their sequential exploitation—thus realizing an equal probability of observing any number in the interval;

inde-pen-dence is hard.

2. the probability of guessing any bit of the next string is *practically* 1/2—thus realizing the mutual independence of the numbers in the sequence.

ATM is sure (enou-gh)

The word *practically* is the keystone of a huge literature on random number generators [3] where the closeness to 1/2 of the above probability deriving from feasible algorithms becomes increasingly stronger. This is why we may manage our bank account electronically without relevant risks.

Algorithm 2.1 Linear congruential generator

Given a, b and m generator-specific integer constants, a linear congruential generator is defined by the recurrence relation:

$$u_{j+1} = \frac{\left(a \quad m \quad u_j + b\right) \bmod m}{m} \tag{2.1}$$

where $\{u_j\}$ is the sequence of random values in $[0, 1]$, u_0 the initial seed and mod is the modulo operation. Typical choice of the parameters is $a = 1664525$, $b = 1013904223$, $m = 2^{32}$.

2.4.3 Entropy

Entropy: the moment m_1 of a very special ran-dom vari-able

The Shannon entropy *H* is a *sui generis* moment of a random variable that constitutes a first connection claimed in the literature between physics and information content of a phenomenon. Its expression, as a function of the probability distribution P over an event space Ω that one may have partitioned into the set $\{A_1, \ldots, A_n\}$ is the following:

$$H(P) = -\sum_{i=1}^{n} \mathrm{P}(A_i) \log \mathrm{P}(A_i)$$

Referred to a random variable X describing the A_is in the role of events, in the continuous release by commodity, it reads:

$$H(X) = -\int f_X(x) \log f_X(x) dx$$

This quantity is the uncontested queen of most topics concerning information management. So, there is a vast literature showing properties and exploitation of it [5]. Here, we just point out that H is a positive function with the general meaning of uncertainty indicator, assuming value 0 in case of certainty, i.e. when P assumes uniquely values 1 and 0. A peculiarity of this function is that it uses the same distribution probability at two levels: both as an argument and as a function of itself. It is a situation analogous to the liar paradox on Sect. 1.3, however, without the same inconvenience because now we do decide not the truth value of an event but its uncertainty. Note that the result is a non-dimensional quantity, since P() is a number without dimension (no length, voltage and so on). It takes the average of the log of the probabilities of the events into which we partitioned the sample space. Thus, the value to be averaged depend on how we decide to *observe* the space (i.e. to partition it). We may relate the certainty degree of the single element A_i of the partition to its probability $\mathrm{P}(A_i)$, so that a complementary measure of uncertainty may be related to its inverse $^1/_{\mathrm{P}(A_i)}$.

$H = 0 \rightarrow$ certainty; $H > 0 \rightarrow$ uncertainty.

Uncertainty is a matter of perspective.

From another perspective, if someone tells me my son has passed a test but I know already, we may drink to it, but I do not get any new information. Vice versa, it this notice reaches me when I'm still not sure of the outcome of the test, I gain information. This the rationale of denoting in Information Theory $\log {}^1/_{p(A_i)}$ (i.e $-\log p(Ai)$) the *information* of the event A_i. The reason for reshaping the probability with its log will be discussed later on Sect. 4.7.1.2. Thus the information entropy introduced by Shannon [12] plays the role of a clear index of uncertainty of our observation of a sample space, and of the quantity of information we need to remove that uncertainty.

Great uncertainty *ex ante* $\rightarrow$ great information gain *ex post.*

A somehow analogous concept of entropy was introduced earlier (about one century before) by Boltzmann as an interpretation from an information perspective of the original entropy S introduced by another physicist, Clausius, a few decades earlier. According to Clausius, the entropy is a relative quantity whose variation may be measured during a thermodynamic transformation via

Entropy is a matter of energy transfer,

$$\frac{\delta Q}{T}$$

where T is the uniform temperature of a closed system and δQ the incremental reversible transfer of heat energy into the system. According to this definition, the Second Law of Thermodynamics establishes that the entropy of an isolated system cannot decrease. This gives a direction to any thermodynamical evolution of this kind of system and, consequently, of the entire Universe in the widest instance.

According to Boltzmann, the absolute value of S is linked to the notion of order of the above system. The formula, concerning the entropy S of a gas is,[34]

but a matter of variety of compatible states, as well.

$$S = k_{\mathrm{B}} \ln W$$

where k_{B} is a dimensional constant, since S has the dimension of an energy, and W is the number of actual microstates corresponding to the gas macrostate. A macrostate is identified by the overall parameters of the gas, such as temperature and pressure; a microstate is the record of the local features of each gas particle determining the

[34] We may read it carved on Boltzmann gravestone.

Hence it is a matter of probabilities, as well.

macrostate as a whole. Under the hypothesis that each microstate μs corresponding to a given macrostate has the same probability $p(\mu s) = 1/W$, $\ln W$ is exactly the entropy H of the macrostate (apart from the constant $\log e$). So that the second law of the thermodynamics decrees the evolution of isolated systems toward increasingly uncertain conditions. In this sense, S seems to be an inner clock of any natural event [4]. Thus, looking at a heap of tree trunks and a heap of cinder in its place, we have no doubt that the former precedes the latter along the timeline.

Apart from a somehow mystic evaluation of the energy of a bit of information [9], the analogy between H and S seems to close the loop from *In the beginning chaos came to be* to a more ordered condition, as if Nature bent partially to the needs of human beings, and goes irreversible back to the chaos. In this, the trip from order to chaos goes hand in hand with the increase in the number of the underlying microstates, from a selected number limited by humans to the final anarchy.

Find the demiurge in you.

The comparative greater richness of the Shannon entropy is the non equiprobability of the microstates, or, equivalently, the exploitation of symmetries on families of them. This allows us to pass from the operationally non-distinguishable microstates of a gas to distinguished families of them in the role of exploitable intermediate states. As we will discuss in next chapters, their exploitation requires systems more complex, such as humans are, than the simple ensembles of particles of statistical mechanics in physics. However, the great merit of the Boltzmann constant k_{B} lies in having rooted the information of a system on its physical features and vice versa. Thanks to this hybridization, we may use the Shannon entropy to monitor the evolution of a (real or pseudo) natural process toward a goal of interest to us [6]. A well-known template of this strategy is represented by *simulated annealing* algorithms [11].

2.4.4 Conditional Probability

We just recall the basic formula. It's a formal way of taking into account ancillary information on a phenomenon in question.

The conditioning event is our current probability space.

Consider a probability space (Ω, Σ, P), for instance describing the car pool in Italy, as for items and related probabilities. If we are concerned with the fuel consumption and in particular with the probability that the next car I encounter (in the ideal experiment, I will just pick randomly in the pool) has a (standard) fuel consumption of less that 0.3 L/km ($X < 0.3$), I must consider the set A of Ω in the picture below. However, If I'm only interested in engine volumes of less than 1.2 L, my sample space becomes the set B in the picture, and the event of interest to me is the intersection $A \cap B$ between A and B. Consequently, the *Conditional Probability* of the event A given that the event B occurred is

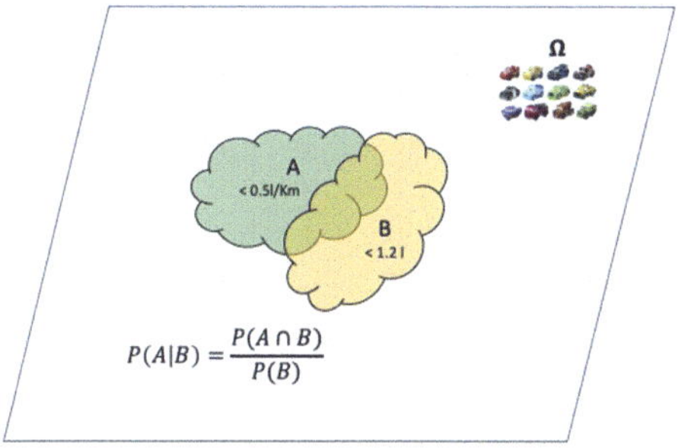

$$P(A|B) = \frac{\mathrm{P}(A \cap B)}{\mathrm{P}(B)}$$

From this basic formula a lot of expressions derive, including the conditional probability density distribution and cumulative distribution of a random variable that we will meet in the book. In particular, if $\mathrm{P}(A \mid B) = \mathrm{P}(A)$ we say that events A and B are stochastically independent—a statement that is exploited in the definition of the likelihood of sample on Sect. 2.1.

Actually, conditional probability constitutes the root of the Bayesian approach to inference. In this book we will mention some particular aspects of this approach; however we skip a systematic introduction of it so as not to excessively broaden the book's scope (remember, we are not aiming for a *Summa Teologica*).

References

1. B. Apolloni, S. Bassis, Confidence about possible explanations. IEEE Trans. Syst. Man Cybern. Part B: Cybern. **41**(6), 1639–1653 (2011)
2. B. Apolloni, S. Bassis, D. Malchiodi, W. Pedrycz, *The Puzzle of Granular Computing*, volume 138 of *Studies in Computational Intelligence* (Springer, Berlin, 2008)
3. O. Goldreich, *Modern Cryptography, Probabilistic Proofs, and Pseudorandomness*. Algorithms and Combinatorics. (Springer, Berlin, 1999)
4. S. Goldstein, J.L. Lebowitz, R. Tumulka, N. Zanghi, *Gibbs and Boltzmann Entropy in Classical and Quantum Mechanics* (WorldScientific, Singapore, 2020), pp. 519–581
5. R.M. Gray, *Entropy and Information Theory* (Springer Publishing Company, Incorporated, 2011)
6. K. Marti, *Stochastic Optimization Methods: Applications in Engineering and Operations Research* (Springer Publishing Company, Incorporated, 2016)
7. A.M. Mood, F.A. Graybill, D.C. Boes, *Introduction to the Theory of Statistics* (McGraw-Hill, New York, 1974)
8. R. Motwani, P. Raghavan, *Randomized Algorithms* (Cambridge University Press, Cambridge, 1995)
9. J.D. Norton, Eaters of the lotus: Landauer's principle and the return of maxwell's demon. Stud. Hist. Philos. Sci. Part B: Stud. Hist. Philos. Mod. Phys. **36**, 375–411 (2005)
10. V.K. Rohatgi, *An Introduction to Probablity Theory and Mathematical Statistics*. Wiley Series in Probability and Mathematical Statistics (Wiley, New York, 1976)
11. C.D.S. Kirkpatrick, Jr. Gelatt, M.P. Vecchi, Optimization by simulated annealing. Science **220**(4598), 671 680 (1983)
12. C.E. Shannon, *The Mathematical Theory of Communication* (The University of Illinois Press, Chicago, 1949)
13. S.S. Wilks, *Mathematical Statistics*. Wiley Publications in Statistics (John Wiley, New York, 1962)

Chapter 3
Data First

Abstract As compatibility is a key issue of this book, the starting point is the observed data. In this chapter we provide a general model of them, as well as a consequent general approach to derive distributions of parameters determining compatible descriptions of them. A description may be delivered either in terms of the data distributions (inferring distributions), or even of the functions underlying the relations between the data components (learning functions, both Boolean and continuous). The common pivot is the magic U, i.e. a hypothetical uniform random variable as the pure primer of the randomness of data and parameters. The magic of U (or other similar seeds) is that we may hypothesize its properties without requiring its real existence, and the inference procedures we derive are very elementary and clear. Nonetheless, they provide an essential framework where our golden kit may operate to attain all inference goals we desire. The framework we provide promises to be capable of jointly dealing with abstract and physical variables, hence those directly generated by neural networks and those by any external actuator, in a compatible world where mental and physical operations share a common objective function.

Till now, we have discussed description and explanation tools that may be suitable for any possible world, any of the multiverses. Let's call these tools *operators*. But, personally, I'm interested in the actual world surrounding me. *Actual* is a big word, which we learned to smoothen into *compatible*. To exploit these operators, we need to establish a link with a compatible world, whose hook on our part is constituted by the data we observe. Let's start from the beginning and focus on data that are representable through scalar (i.e. one-dimensional) variables. For sure the surrounding world is too complex to be represented by such simple variables, but learning to manage them is a good starting point and enough for this book.

In search of a link with, at least one-dimensionally, observed variables.

B. Apolloni, *A Few Things I Know About Her*, Intelligent Systems Reference Library 219, https://doi.org/10.1007/978-3-030-94379-0_3

3.1 An Essential Cosmogony of Data

Let's start from our elementary and fully symmetric scenario constituted by a set of elementary items, which are affected by a feature that has the same value on each (exactly for symmetry reasons). We are used to representing this set through a two-dimensional stick. In principle, it is the blue rectangle in the left part of the figure below, where, along the length (the x-axis) we locate the items uniformly—each item occupying a same width with neither overlaps nor holes. In such a way they take a progressive allotment number that represents the item index or its position normalized in [0, 1]. The height is equal as well, representing the questioned feature.

Stretching the unitary probability mass.

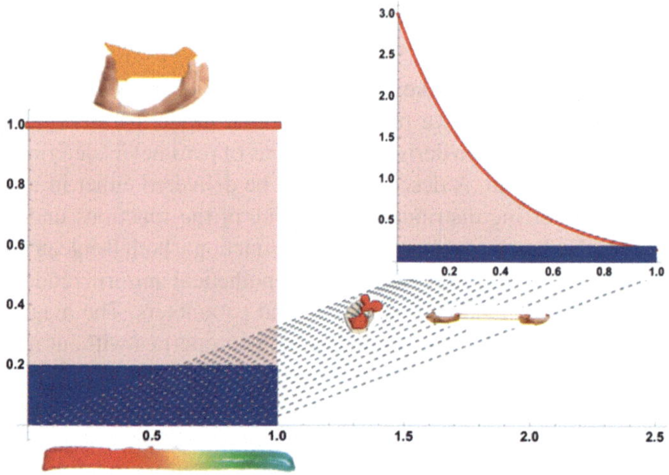

Within the description thread, we may use the continuity facility to identify the set of items with the points of a unitary segment, naturally ordered from left to right for computational reasons (remember your mom asking you to put your room in order). Within this continuous representation, the red rectangle over the bar represents the uniform density function into which the equal probabilities of the items are smeared. The target feature is the density function, so that the early model is a uniform random variable U spanning the interval [0, 1]. Now, imagine the stick to be gummy, like the jelly candy reported under the stick, so that we may variously squeeze (or expand) it along the x-axis, with the results that now, to maintain the overall mass, the eight of the stick is no longer uniform. If we are well skilled, it may assume a desired shape. In the above picture, starting from the uniform density function of U we arrive to a density function of the Exponential distribution law like the one on Sect. 2.3.1.1. The colored gummy bar evokes the squeezing strength exerted on the bar, compression strength (red) on the left, expansion one on the right (green) with an intensity growing from the center towards the extremes of the bar.

We squeeze the x-axis and modify the y-axis, consequently.

Compressing/expanding may be done in many ways. The most common ones are linked to the shape of the CDF of the target distribution. This derives from two factors:

F_X as a formidable molder.

1. the monotonicity of the CDF which renders univocal the solution of the inverse problem: which is x such that $F_X(x) = \alpha$ for a given α (see the definition of quantile on Sect. 2.4.1.1)
2. the universality of the distribution function of the variable $F_X(X)$ (see the introduction to a function of a random variable on Sect. 2.4.1.1)

$F_X(X)$ is uniformly distributed on the y-axis.

At the beginning a huge set of equal frequency data,

Namely, the Probability Integral Transformation Theorem [24] recognizes that the result of applying the cumulative distribution function F_X of a continuous random variable X exactly to X is a uniform random variable U in $[0, 1]$ (obvious variant for X discrete). The claim should not prove astonishing if we recall the role of F_X. Namely: $F_X(x) = P(X \leq x)$, which we may read as the asymptotic frequency of the number of items below the threshold x in an infinitely large sample. To obtain the approximation of F_X on a finite sample $\{x_1, \ldots, x_m\}$, let us call it $\hat{F}_X$ we may sort the sample as $\{x_{[1]}, \ldots, x_{[m]}\}$ (with $x_{[i]} < x_{[i+1]}$) and get the points $(x_{[i]}, \hat{F}_X(x_{[i]}) = (x_{[i]}, i/m)$, with obvious filling between consecutive items, so that $\hat{F}_X(x) = \hat{F}_X(x_{[i]})$, for $x_{[i]} \leq x < x_{[i+1]}$. In this way, we see from the picture below that $\hat{F}_X(x_{[i]})$s are uniformly spread in the interval $[0, 1)$ and, since each of them, represented on the y-axis, corresponds to a sample item on the x-axis drawn with equal probability $1/m$, also $F_X(x_{[i]})$s enjoy an equal probability $1/m$, thus representing the values of the discrete random variable U uniformly distributed in $[0, 1]$—which asymptotically converges toward the continuous companion U. In other words, passing from a huge sample of Uniform distribution to a huge sample of any other random variable, what changes is the location of the points in the x axis, not in the F_X axis.[1] The passage from U to X is operated by the function F_X^{-1} as shown in the right part of the picture below. Namely, the blue curve represents the function F_X mapping from X to U, as the results of infinite applications of the pairing $(x_{[i]}, i/m)$. In the opposite direction, crossing any horizontal line of height u on the vertical axis with this curve, we obtain the corresponding $x = F_X^{-1}(u)$.

[1] Note that, in the case of discrete random variables, the motions on the x axis could lead two or more $x_{[i]}$ to collapse, and this generates jumps of different height on the F_X with respect to the x values.

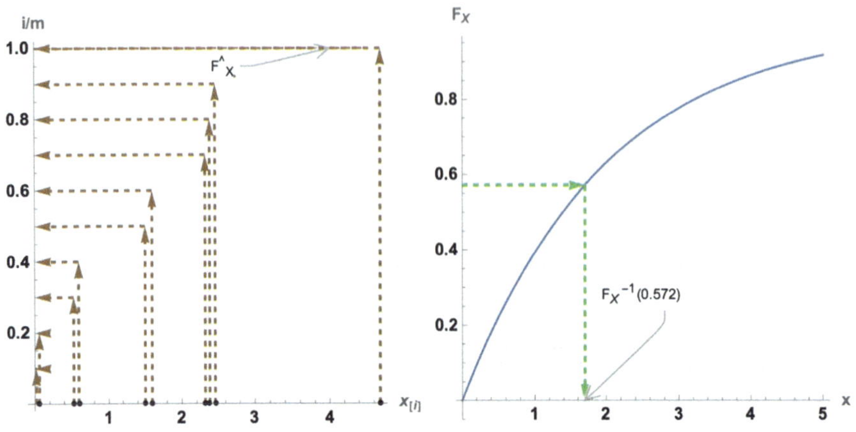

A bit more formally: $F_X(x)$ defines the *inverse* function F_X^{-1} so that $F_X^{-1}(F_x(x)) = x$ (For instance the inverse of the multiplication by a is the division by a; indeed $1/a \times ax = x$). Now, if we apply F_X to the random variable X, we obtain a random variable $Y = F_X(X)$, so that with $F_Y(y)$ we measure the probability of the event $Y \leq y$ that we may rewrite as $F_X(X) \leq y$, i.e. the event $X \leq F_X^{-1}(y)$ whose probability is measured by definition of the CDF by $F_X(F_X^{-1}(y))$, which in turn is equal to y by definition of inverse function.

In conclusion, for any continuous random variable X, we have $F_Y(y) = y$, which denotes Y to be a uniform random variable in [0, 1]. The same holds for discrete variables as well, where F_X^{-1} generalizes as $F_X^{-1}(u) = \min\{x | F_X(x) \geq u\}$.

3.1.1 Sampling Mechanism

then a sampling mechanism to generate the desired variable, via the Universal Sampling Mechanism.

We may use the above claim to devise a data generator as in the figure below, which we will assume as the *mother* of all samples. We may see the horn in its center as a spring generating uniform random numbers in [0, 1] for free, which are then transmuted into numbers following a given distribution law after having passed through the gears of the mechanism. Let's denote it by g_θ. Thus, we have a *universal sampling mechanism* $\mathscr{M}_X^*(U, g_\theta)$, such that for each sample u_i of U we obtain a sample $x_i = g_\theta(u_i)$ of X, and we are ensured that, like from a cornucopia, you may draw any kind of random variable X, provided we may devise the gears in order to compute the related inverse function F_X^{-1}. This is shown in the picture below.

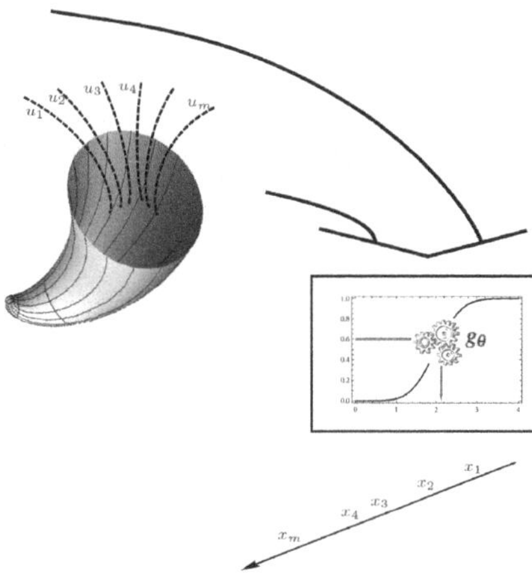

More in general, we may consider a *sampling mechanism* $\mathscr{M}_X$ constituted by the pair (Z, g_θ), where Z is a completely specified random variable and g_θ is a function that maps from values of Z to values of X, such that for each sample z_i of Z we obtain a sample $x_i = g_\theta(z_i)$ of X. We call g_θ the *explaining function* and focus on it to describe X.[2] Of this function, g represents a general shape that we may suppose derives from previous experience or an abstract model, θ denotes a set of parameters, the parameter as a whole, which specializes g to generate exactly values of X.

But we may have other Sampling Mechanisms.

In this way, the mentioned dichotomy description-explanation—and related statistics and physics frameworks—tends to vanish. In both cases we are looking for an explaining function of the conversion of a seed into a target variable. Normally the seeds in statistics are random variables, while in physics they are specific values. But we will see that this difference too will become less sharp in complex tasks such as regression and learning. Hence, *explaining* will be the leitmotif of this book, where the general concept of sampling mechanism do override those introduced on Sects. 2.3.1.1 and 2.3.2, and will be variously implemented in the following. *Sample* itself becomes the general notation of any observation (either individually or as a set) of a phenomenon, where, if we do observe, it is because something is not known and this lack of knowledge is recovered by some random source.

3.1.2 Summing Up

Our trip starts from a set of data produced by Nature. Our favorite way of approaching the data is to find an early organization simply by sorting them. It is a non trivial task

---

[2] As for an example of a different explaining function linking X to the seed, we may map a generic Gaussian variable Y from the standard one Z through the formula

$$Y = \mu + \sigma Z.$$

requiring us to characterize the data with comparable features x_is so that it makes sense to compare their values, i.e. check whether $x_i \leq x_j$. If this check is feasible for any pair (i, j) spanning the entire set of the n data, we may draw an initial synthetic view through the Empirical Cumulative Distribution Function (ECDF) which reckons on any value x of its argument the number of data items less than or equal to x. Denoting this function as $\hat{F}_X(x)$, its definition is

In a sample of equivalent observations, one is worth one,

$$\hat{F}_X(x) = \frac{\text{number of } x_i s \text{ such that } x_i \leq x}{\text{number of } x_i s},$$

i.e., recalling the $F_X(x)$ definition on Sect. 2.4.1.1,

$$\hat{F}_X(x) = \begin{cases} 0 & \text{if } x < x_{min} \\ \sum_{x_i \leq x} 1/n & \\ 1 & x > x_{max}. \end{cases}$$

In other words, we reckon a frequency $\phi = 1/n$ on each data item, and the curve $Cum\phi$ accumulate the frequencies along the sorted data. We assume the latter to represent a *sample*, hence to be provided by nature without any bias. We may see the samples to be the customary expression of the Nature so that those provided today will not differ from those she will provided tomorrow; rather, if they change it means that Nature has changed her habits. Moreover, when these data are so many, we may identify them with the entire *population* of data that Nature will provide in her eternal live (as long as she doesn't change habits). This allows us to interpret this instantiation of ECDF (based on frequencies) as a CDF (based on probabilities), where the shift from constant frequencies to possibly non constant probabilities is accompanied by the change of their argument: frequency $1/n$ refers to the single data, probabilities to the values they assume. This is for the right part of the picture below.

better if organized through $\hat{F}_X$,

to get a model F_X.

CDF on the right is the model of the Nature, as for her source of expressed data. We may manage it in various ways, for instance moving from CDF to PDF, as in the left part of the picture, for instance moving to discrete values that could be of our interest (the yellow graph). Once the model is fixed, we may assume further data, provided by Nature for instance in a definitely smaller number m, as generated by this model. Our entry-level compatibility issue is that the model shape is at the basis of our personal multiverses, in the sense that we know it except for a parameter ψ (in turn, possibly constituted by a set of parameters) which may specify the universe that actually produced the data.

If we had a model we could generate data like Nature.

If we lack a parameter of the model,

The lower part of the figure is our way of solving the issue. We pair up the sample of data provided by Nature with a sample of a uniform random variable provided by us,[3] where the link between the two samples is established by the explaining function g_θ. In this way, the parameter ψ of the model maps into the parameter θ of the explaining function, whose identification is the core business of the *inferential* methods.

we must infer it by delving into the sampling mechanism and observations.

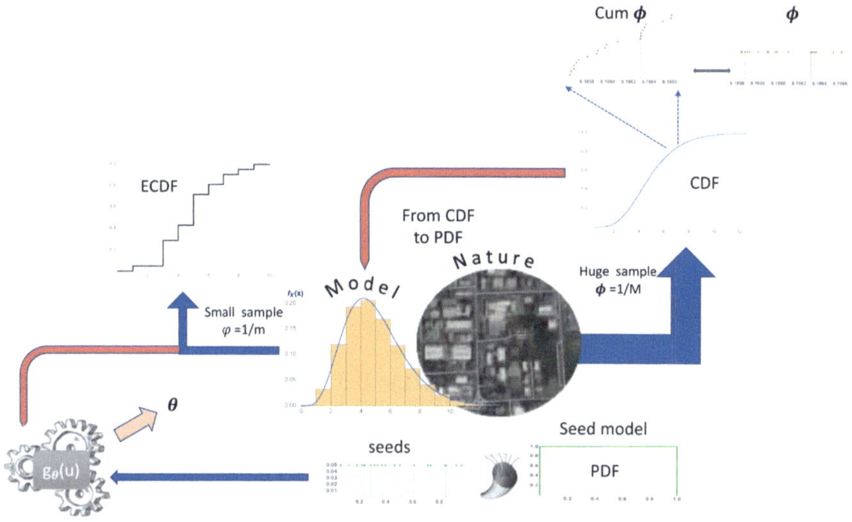

3.2 The Inferential Problem

Let's come back to our unitary stick and assume that someone, Bob, used $\mathscr{M}_X^*(U, g_\theta)$ to squeeze/expand it. This means that he used a map on x-axis as the following

$$x = F_X^{-1}(u)$$

Consequently, the slope of the explaining function $g_\theta = F_X^{-1}$ increases where the distance between the mapped points increases and vice versa where it decreases.

Aiming to update the description of the stick we should know g_θ. And, since Bob doesn't want to reveal it, we must discover it simply by observing the deformed stick. *Observing* is another dedicated word. In our case, the goal is to resume some relevant features of the deformed shape, which may prove useful in the future. While our eyes are able to *capture* such features very often, because they are supported by our brain, if we want to commit an analogous task to a computer (maybe configured exactly

A feasible way of guessing g_θ,

[3] Please, don't be so peaky as to say that a generator like the congruential one on Sect. 2.4.2 is a *pseudo*-random generator. It's as random as it needs to be, and if it is non sufficient then we will use a stronger one.

as an artificial neural network), the features should be the output of an algorithm fed by *all* the points in the stick and their heights. But these points are in an infinite number.[4] Thus we must look for shortcuts.

- basing on samples

- A former is the sampling. Rather than considering all points, we pick here and there points x_is on the x-axis, so as to get a sample with the properties discussed on Sect. 2.1, which is *representative* of the entire x-axis.
- Since the inversion problem: what is the explaining function that could have generated the sampling via the mechanism

$$x_i = g_\theta(u_i)$$

remains a difficult problem (because of the shape of the function) and indeterminate (because the non observability of the originals u_is), a hard softening occurs in two steps:

-focusing on θ

- coming to compatible θs

1. assuming g_θ to be known as for the general shape—known g—so as to focus the question on the sole value of θ
2. looking for *compatible* values of θ[5]

This is what we discussed in relation to the Bernoulli variable on Sect. 2.1.1 and ff.

A unique framework for bodies, gases and random data.

Looking at the red feature in the shooter figure on Sect. 3.1 as a graph of physical strengths, we may translate the previous discussion in terms of mechanics of elastic-incompressible bodies or states of an ideal gas Both these systems, and the previous description system as well, share a principle of conservation of a quantity that may be the internal energy for gases, volume for bodies and *existence* for description. Within this conservation we are allowed for infinite degrees of freedom that determine the feature landscape. This calls for inference procedures, where the questioned feature plays a dual role:

A dual role of the features: inner engine and external interface.

1. an inner one, as the result of a constrained equilibrium between contiguous points,
2. an outer one as the primer of the observables in input to our statistics.

Distributions too are the outcome of a process.

As for the former, the picture below shows the evolution of a sample of 20 points drawn from a rather symmetric distribution and from an asymmetric one, respectively, and endowed with an initial equal probability 1/20 smeared in non overlapping rectangles (width × height equal 1/20) around the points. The evolution is ruled by atomic interactions between sub-rectangles of the above distribution, aimed at establishing an equilibrium confirming the symmetry and non symmetry of the source distribution.

[4] Still worse, mathematicians point out that the stick is made up of an infinitely non numerable set of points.

[5] By default we will refer to scalar parameters as well.

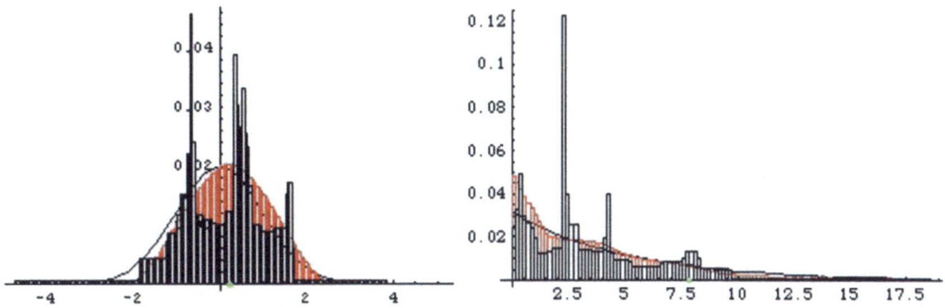

More in detail, the experiment is the following

- Draw randomly 20 observations (a sample) from a rather symmetric phenomenon and from an asymmetric one, respectively, like the red points in the figure below.
- attribute equal probability $1/20$ to each point, because of their generation, and smear this value in non overlapping rectangles (width $\times$ height equal $1/20$) around the points. This depicts the mentioned indeterminacy of the discrete operations. We drew, for instance, points 0.3, 0.7 and 0.9, we may realize that a next point 0.42 is preceded by one sampled point and followed by two sampled points, hence is rather low in that framework. But we cannot make serious distinction between point 0.42 and point 0.56 on this basis. This is why we give the same probability density inside the above rectangle, which we center on the sampled point for minor reasons.

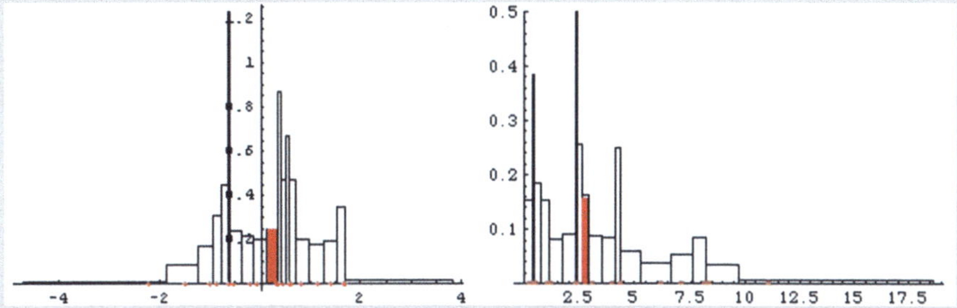

- Raster the abscissas according to a discretization step equal $1/50$, getting the pictures below

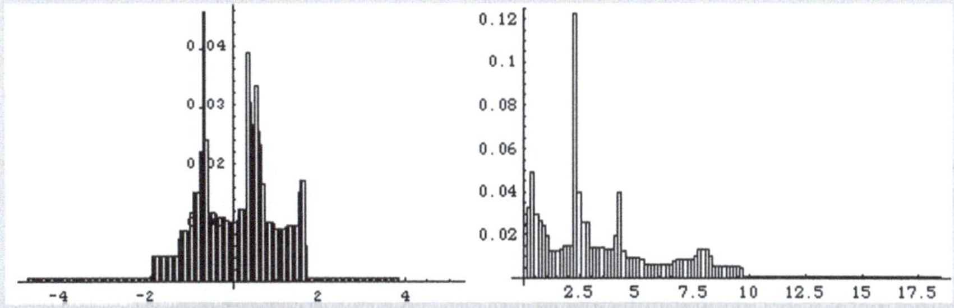

- On the left picture we iteratively force the following conditions

  1. the probabilities must be symmetric around a centroid (highlighted by the red bar in the previous picture),

2. the probabilities must be monotonically decreasing on the right and increasing on the left of the centroid

by simply flushing the graph with a wave of probability atoms of size ε from left to right and vice versa. More precisely, along one direction we add ε to the vertical bar if

- either it is lower than its symmetric counterpart or
- it is lower than the next bar if we are on the right of the centroid, or of the previous bar if we are on the left.

Along the opposite direction we subtract ε to a bar in complementary conditions.

- On the right picture we iteratively force the following conditions

1. only positive values are allowed,
2. the highest value of the probabilities is in the origin, and
3. the course of the probabilities is a monotonically decreasing slope

The operational procedure is the same same as above, but implemented in the direction from 0 on.

Statistics are properties of random samples.

As for the second aspect, *statistics* are the tools through which the *microstates*[6] of a system interact with us.

A statistic is any function of the observed variables, such as the output of any routine having on input those observations. Thus, in an ideal gas, the molecule dynamics determine the inner pressure (which is constant in equilibrium conditions), while the pressure on a given wall is a statistic on the gas, since it sums up to the force exerted on this wall. Actually, it is a peculiar statistic since it's a function of variables that we actually do not observe unless through this pressure.

In our squeezable stick, the explaining function determines the probability distribution. In turn, this distribution determines the moments, as defined in the point 4 of Sect. 2.3.1.1, or other properties of interest to us that are observed through companion functions, the statistics, applied to samples of the distribution in question. Things are a bit more intriguing in the frameworks of mechanics, where the compression/stretching forces act on the stick to change its shape, and are in general the target of our observation, hence the source of interaction with us as well.

Our strategy is to transfer probabilities from random statistics to random parameters.

As for the focus on probability distributions, statistics are the tools to identify the parameter θ of the explaining function, which results in a *statistical inference*. As noted in the case study of the football wins, since we base our computations on sample data, we cannot expect to identify *the* value of θ. Rather, basing ourselves on much informative samples and much informative statistics, we may aim to compute the probability distribution Θ that describes the compatibility of its values with the sample. This is the goal of an inference on θ.

To render the sample very informative, we require by default that it has been drawn without any bias (like the numbers of a honest lottery), so that it may be considered to be constituted by a specification of a set of random variables having the same g_θ (hence the same distribution) and being stochastically independent of any another.

[6] See Sect. 2.4.3 for further discussion about them.

To work with much informative statistics, we require the latter to be well suited for the inference procedure below. Namely, for a general sampling mechanism $x_i = g_\theta(z_i)$ as on Sect. 3.1.1, identifying θ from a sample requires to solve a system of equations

Inverting the sampling mechanism on each item may prove nonsense,

$$\begin{pmatrix} x_1 = & g_\theta(z_1) \\ \dots & \dots \\ x_m = & g_\theta(z_m) \end{pmatrix}$$

that may be solved in terms of an infinite (non numerable) set of pairs $(\theta, \{z_1, \dots, z_m\})$ since the seeds are unknown as well. Thus we make the problem less indeterminate by focusing on a relevant property of the sample $s = \rho(x_1, \dots, x_m)$. Namely, if we replace the x_is with their expression $g_\theta(z_i)$ in ρ we obtain the following expression which we denote as *master equation*[7]

inverting on a property of the sample may be feasible and effective,

$$s = h(\theta, z_1, \dots, z_m) \qquad \text{(Mastereq)}$$

The relevance of the statistic is related to the ability to have one and only one solution in θ to the above equation for whatever s and $\{z_1, \dots, z_m\}$, so as to insure a univocal transfer of probability from seeds $\{z_i\}$ to a specification of θ for any s. This requires a *well-behaving* ρ that we may identify with these properties:

provided we rely on a well-done statistic.

1. *strongly monotonic relationship.* A uniform monotonic relation exists between s and θ for any fixed seeds $\{z_1, \dots, z_m\}$;
2. *well-definition.* On each observed s the statistic is well defined for every value of θ, i.e. any sample realization $\{x_1, \dots, x_m\}$ such that $\rho(x_1, \dots, x_m) = s$ has a probability density different from 0;
3. *local sufficiency.* The probability of observing the sample (i.e. the product of the PDF of its elements) depends on θ only through the computed s, so that for a given s it does not depend on θ.

Three conditions for a proper probability transfer,

Relying on this statistic, we can generalize the method exploited with reference to the Bernoulli variable in the Algorithmic Inference framework[8] on Sect. 2.1.1.1 to an *inference procedure* made up of four steps

and a general inference procedure.

[7] As mentioned before, master equation is a rather generic notation. In this book we use it to denote the pivot of some inferential methods. Hence we will encounter some further instantiations of it throughout the text.

[8] To avoid doubts, Algorithmic Inference is a general framework where statistical inferences are carried out.

The general Algorithmic Inference procedure

1. *Selecting the family of explaining functions.* It depends on what is known about the random variable X under question. We remark that besides $g_\theta = F_X^{-1}$, we may have many different explaining functions linking X to the seed.
2. *Identifying the relevant statistic.* i.e. a statistic well-behaving, if any, or a surrogate of it, if necessary.
3. *Establishing the master equation* linking the statistics s to the parameters θ for whatever sample $\{z_1, \ldots, z_m\}$ of the seed.
4. *Solving the master equation* in θ.
5. *Defining the distribution law* of Θ. Note: we lack a θ value and obtain a random parameter Θ.

Point (2) is the key to the Neyman approach to inference [18]. In that framework, the queen of statistics is the *sufficient statistics* S, where sufficiency depends on the fact that the joint probability density function $f_{\mathbf{X}_\theta|s}(x_1, \ldots, x_m)$ of a sample $\mathbf{x} = \{x_1, \ldots, x_m\}$ (i.e. the product of the instantiations of the function f_X on the values x_i constituting the sample) given the value of the statistic is independent of θ. The expression of $f_{\mathbf{X}|s}$ may be resumed from the definition of the conditional distribution introduced on Sect. 2.4.4. The rationale is that if we know in advance that a sample will determine a value of s of S, we may completely compute the probability of observing the actual sample within the set of those with such property even if we ignore the value of θ parametrizing the distribution of X. Thus, sufficiency is a property of S with respect to a parameter θ. For instance that the statistic $\hat{x} = \sum_{i=1}^m x_i/m$ is a sufficient statistic with respect to the parameter $E[X]$ of a Gaussian random variable.

The relevance of the statistic T we are looking for in our procedure derives from its suitability to invert the master equation introduced in point (3). It's interesting to note that sufficiency in the Neyman acceptation implies relevance in ours, which in turn provides a rationale for the property of sufficiency.

Point (5) may be variously carried out. The backbone is the twisting argument instantiated in the above case study of the Bernoulli variable, where the horizontal feature of the threshold line on Sect. 2.1.1 translates into the invariance of g_θ affecting both past and future observations. For instance, an immediate extension is shown in the following example.

Example The Negative Exponential random variable X with parameter λ, has a CDF

$$F_X(x) = 1 - e^{-\lambda x}$$

So an explaining function for it is

$$g_\lambda(u) = \frac{-\log(1-u)}{\lambda}$$

which denotes statistic $s_\lambda = \sum_{i=1}^m x_i$ to be well behaving. The corresponding master equation reads

$$s = -\frac{\sum \log(1-u_i)}{\lambda}$$

which provides the solution

$$\lambda = -\sum \frac{\log(1-u_i)}{s}$$

For this parameter we may establish the (Logic)

$$\left(\lambda \leq \widetilde{\lambda}\right) \Leftrightarrow \left(s \geq s_{\widetilde{\lambda}}\right)$$

where $s_{\widetilde{\lambda}} = \sum_{i=1}^m \widetilde{x}_i$ and $\widetilde{x}_i$ is the value into which u_i would map substituting λ with $\widetilde{\lambda}$ in the explaining function. Proceeding in analogy to Sect. 2.1.1, we come to the (*Metaprob*) equation

$$F_{\Lambda_s}(\lambda) = F_{S_\lambda}(s)$$

where the relation is exactly an equation since the involved variables are continuous, so that no injection of indeterminacy occurs. From this equation we derive that the random parameter Λ follows the distribution Gamma with parameters (s, m), where m is the size of the observed sample.[9] The pictures below show a sample of an exponential distribution through its ECDF and the corresponding distribution of Λ taking values λ's.

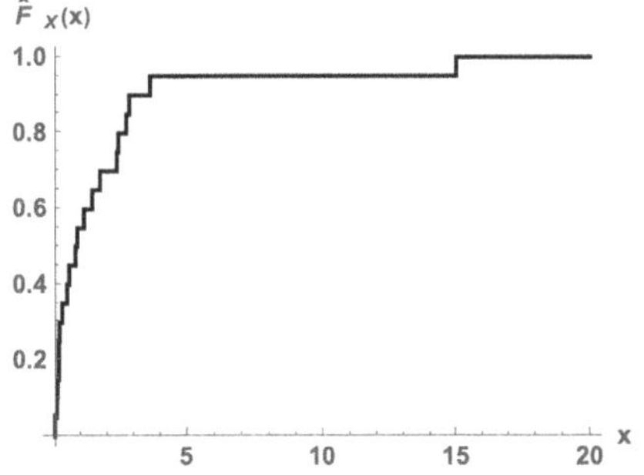

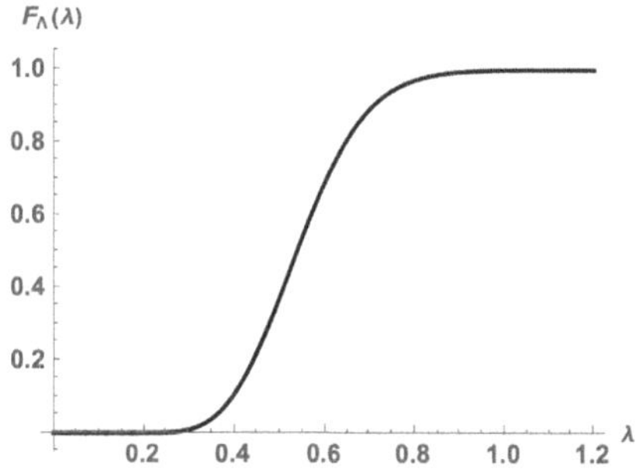

When the analytical form of F_Θ is not easily available, we may use numerical solutions of the master equation do derive directly the empirical distribution of the parameter. For instance, from a set of 1.000 solutions of the above master equation for λ we draw in red the ECDF $\hat{F}_\Lambda$ of Λ in the figure below, that overlaps almost perfectly with the F_Λ (in black) computed above in terms of Gamma distribution.

9

$$F_\Lambda(\lambda) = \left(1 - \sum_{i=0}^{m-1} \frac{(\lambda s)^i \, \mathrm{e}^{-\lambda s}}{i!}\right) I_{(0,+\infty)}(\lambda)$$

where the roles of variable specification and parameter commute between s and λ.

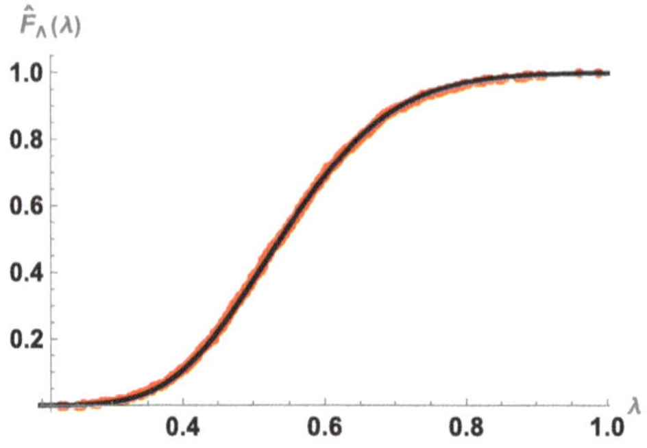

3.3 Compatible Functions

Separating physical concerns from random ones, when possible;

Compatibility is not in the mood of classical physic and its exploitation by engineers. Certainly, the scrap load of an iron rod is a non fixed value, and we need to take into account its randomness in the design of a reinforced concrete structure. However, we may deal separately with the the two problems, randomness of the scrap load and design of the structure for a given scrap load. Consequently, moments m_r's on Sect. 2.3.1.1-point 4 are of great interest, while the companion M_r's are of a scarce exploitation. The scrap load is the integral of the force density on all points of the rod section, not of sampled ones. Moreover, we may measure the scrap load of the entire section through well assessed scrap test methods and instruments. Measuring the same on single points or their surroundings is more difficult. Rather, we usually affect the former measure by corrective coefficients to take into account its randomness.

with complex problems, it may prove impossible.

However, if the complexity of the tackled phenomenon increases significantly, this factorization of the problem may prove unfeasible.

3.3.1 *The Discrete Version: Classification*

Let's consider, for instance, the task of checking the integrity of the wheel-axle of a railroad car. A common tool to get it is the analysis of UT (ultrasound) signals achieved through boroprobes inserted in the axle cavity. Some features of these signals are considered to be relevant and we need a function h of them separating intact shafts from cracked ones.

3.3.1.1 The Learning Framework

Three strongholds:

Let's denote X the set of the features (maybe, organized as an array) that are observable in an experiment like the axle checking and the boolean variable L assuming value 1, if the shaft is intact, 0, otherwise (the label of the shaft). The difference between the explaining function g_θ and h is that the former maps from unknown random seeds $\{U_1, \ldots, U_m\}$ to the observable variables $\{X_1, \ldots, X_m\}$, the latter from the observable variables $\{X_1, \ldots, X_m\}$ to their labels $\{L_1, \ldots, L_m\}$. Again, we are looking for *compatible* functions, where compatibility is now rooted on the observation of both input and output of the functions. By extending to this instance the problem formulation adopted for the explaining function on Sect. 3.2, we may assume h to belong to a given class $\mathscr{H}$ of functions, inside which it is identified by a parameter ψ (we could denote it as h_ψ to highlight this parametrization). Recalling the three "p"s detailed on Sect. 2.1.1.1, let us consider the paradigmatic example in the picture below and recognize the instantiation of those "p"s in this case. In red we see an element c (concept) of the class of function $\mathscr{C}$ of circles. To this aim, we have available only labeled sample s, consisting of 20 points distributed on the plain with a density function f_X and labeled by a circle or a square, depending on the fact that the point falls inside or outside c. In green we have the hypothesis h we may draw with the directives that it must belong to the class $\mathscr{H}$ of circles (hence $\mathscr{H} = \mathscr{C}$ in this case) and must no contradict the labels of the points: hence points belonging to c must belong to h as well, analogously for those not belonging to c (hence a consistency criterion that is seldom adopted in modern-day politics). The goal is to minimize the probability p of the symmetric difference $c \div h$ between c and h, i.e. the probability of having a point in the region whose points either belong to c but not to h, thus my hypothesis erroneously considers this point as negative ($L = 0$), or belong to h, but not to c, so that my hypothesis faulty considers this point as positive ($L = 1$). Both conditions denote an error, whose overall probability is the analogous of p in the first line of the box on Sect. 2.1.1.1. Therein we have realized that we may deal with it in terms of the random parameter P. Here we have the same need because we know our hypothesis, but not c. Rather, we know that there is an entire population C of compatible concepts. In place of focusing on their distribution, we focus directly on the distribution of $P(C \div h)$, let's denote it by $U_{C \div h}$. Namely, for an individual c we want that the probability on incurring a mistake is low, i.e. $p(c \div h) < \varepsilon$ for any fixed ε. But this probability may be expressed only in terms of values that are compatible with the observed sample, hence we divert to the random distribution of it, which derives from the compatibility distribution of C. So we look for the event $P(C \div h) < \varepsilon$ and ask for a good confidence on it, i.e. the probability of this event to be high enough, say $\mathbb{P}(P(C \div h) > 1 - \delta$ for low values of δ (a metaprobability, again).

- examples of how the function must compute
- a class of functions within which to formulate our hypothesis
- an algorithm computing the hypothesis

Three probabilities:
+ the measure of the error
+ the description of its random course
+the probability of events on the random course

One goal: getting a small value of the error probability with a great confidence.

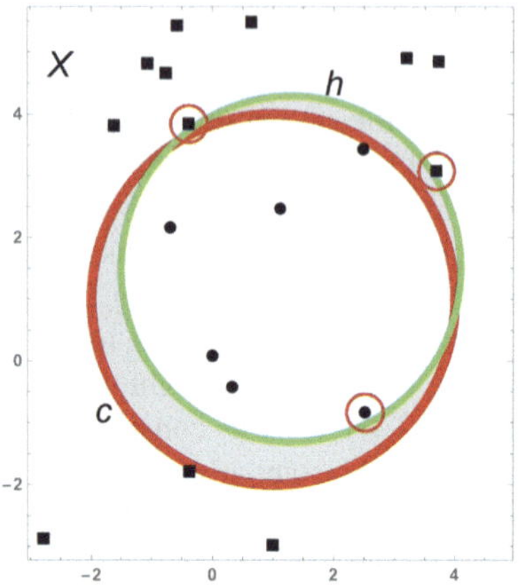

3.3.1.2 The Master Equations

Who states the hypothesis rules?

Resuming, we observe a labeled set $\{(x_1, l_1) \ldots, (x_m, l_m)\}$ and assume that the labels are the output of a function c, namely $l_i = c(x_i)$, where c belongs to a class of functions $\mathscr{C}$, the family of circles in our example. We don't know which one exactly c is, but we know exactly the hypothesis h we draw in its place. We establish h to belong to the class of circles, since we know $\mathscr{C}$ (more in general h belongs to the class of hypotheses $\mathscr{H}$), and to be a consistent hypothesis such that $h(x_i) = c(x_i)$ on all observed points. This delineates a strategy to *learn* c. The result is a distribution law of the probability $U_{C \div h}$ of proving wrong in a next inquiry to h that is no different from the distribution $\mathscr{D}$ we found on Sect. 2.1.2 of the parameter P of a Bernoulli variable, apart from the parameters of $\mathscr{D}$.

If they depend on sniffs on their application instance, then we loose degrees of freedom.

Actually, we could define a Bernoulli variable that takes value 0 or 1, depending on the events "answer correct $(h(x) = c(x))$" or "answer wrong $(h(x) \neq c(x))$", respectively. What changes is the degree of freedom of the sample. We may capture this difference considering an impromptu football match between kids. The result of the match may change if the players state the football rules *before* starting to play, or one of the two contenders is allowed to state special rules *during* the match. It's an elementary consideration that often arises when parliament tries to establish *ad personam* laws. In our example, we draw the circle using the three marked points (two positive points and one negative point), let's call them *sentry points*, that prevent the hypothesis to increase the difference $h - c$ (the positive points) and the difference $c - h$ (the negative point). This changes the equation (Logic) on Sect. 2.1.1.1 as follows:

$$(k_\varepsilon > k + d) \Rightarrow (p < \varepsilon) \Rightarrow (k_\varepsilon \geq k + 1) \quad \text{(Logic1)}$$

The commitment of no cheating on the facts.

- For $k = 0$, it says that, since we opted for non contradicting with our labels (h(x)) the ones we observed (c(x)), we are sure that the probability of $c \div h$ is less than ε if any expansion of it up to a point of reaching a probability measure ε contains all the sentry points that prevent this increase—to be more precise, a minimal set of points that are capable of preventing any *expansion* of $c \div h$ *totally including* of $c \div h$. Thus, imagine we are a dressmaking factory and want to learn about the two main measurements (height and waistline) of our customers. As an early approximation, we may gather them into a rectangle, like in the picture below, where we don't care for the moment about the lower-bounds to these measurements, so that the left-lower vertex of the rectangle is (0, 0), by definition. Hence we have:

1. a class of hypotheses $\mathscr{H}$ made up of all rectangles with positive vertexes, one of them being (0, 0);
2. a labeled sample with positive and negative points;
3. a class of concepts $\mathscr{C}$.

If $\mathscr{C} = \mathscr{H}$, this framework may be represented by the left picture in the figure; otherwise, by the right picture.

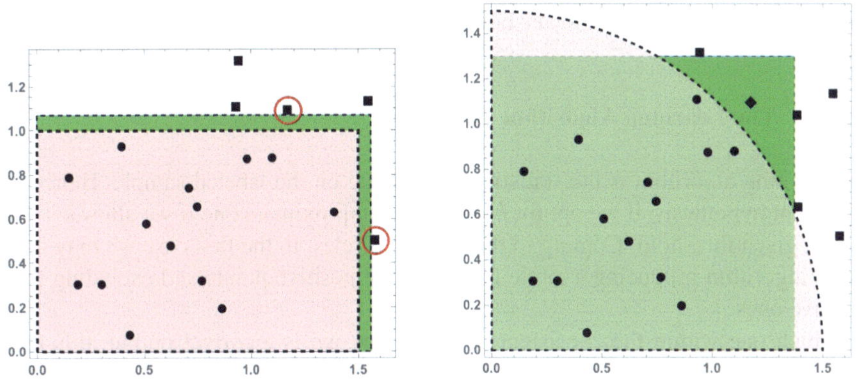

Let's contain the reduction in degrees of freedom, via sentry points,

In the left picture, both the concept (in light red) and the hypothesis (in light green) belong to the class of rectangles. Looking for the largest consistent hypothesis we were prevented from further enlarging the one in the picture by the two marked rectangles: any enlargement requires trespassing on at least one of the negative points. Hence, if a region D_ε of probability measure ε contains both the green region in the picture (namely $c \div h$ here coinciding with $h - c$) and the two sentry points, then also the green region has a measure less than ε, where the sentry points certify that our $h - c$ has not trespassed on D_ε. Hence, the event: two among the points of the labeled sample—the former met by any expansion of $c \div h$—are included in a region D_ε as before implies the vent that $c \div h$ too is included in that region; consequently, its probability measure is $\leq \varepsilon$. In turn, starting from a $c \div h$ with measure $< \varepsilon$ we expand to a greater region by including at least one sentry point, which implies that the number of the items of the labeled sample included in a region of measure ε must be ≥ 1. Put $d = 2$ and we obtain (Logic1).

- In the case of non consistent hypothesis another parameter comes into play, which is the number of mistakes, i.e. of points that we voluntarily include in our hypothesis, even though we know their label to be 0 and vice versa for mistaken positive points. In the right picture, the class of concepts $\mathscr{C}$ is the family of circles centered in (0, 0)

with allowance for some mistakes.

(actually their intersections with the first quadrant). Nevertheless, we maintained the class of rectangles as our $\mathscr{H}$. The hypothesis we draw is a good compromise between accuracy and width of h. This entails, however, that we have inserted a point in the symmetric difference that increases by 1 the number of points in D_ε. In general this number increases by k if this is the number of mistaken points.

So more the sentry points, so less the confidence;

As a result, we may update the equation (Metaprob) on Sect. 2.1.1.1 as follows:

$$\mathrm{P}(K_\varepsilon > k + d) \leq \mathbb{P}(P < \varepsilon) \leq \mathrm{P}(K_\varepsilon \geq k + 1) \qquad \text{(Metaprob1)}$$

mistakes, too, reduce confidence.

Note that, when we opt for a consistent hypothesis no sampled point falls in the symmetric difference, hence $k = 0$. However, we could prefer using non consistent hypotheses, because they are more simple to compute. In this case k represents the number of sampled points mistaken by our hypothesis. It adds to the number d of employed sentry points, representing as a whole an unfavorable parameter that diminishes the probability's lower-bound.

3.3.1.3 The Learning Algorithm

The learning algorithm is the statistic we compute on the labeled sample. Hence a consistent hypothesis, if we opt for $k = 0$, or an approximate one if we allow $k > 0$ up to a given threshold. Coming to the class of circles, in the first case we may ask for an algorithm producing a circle including the positive points and excluding the negative ones.

If consistent, any hypothesis is OK.

Which one within the consistent circles? Any one is a correct output; this the *indeterminacy* price we pay to the fact that we are working with discrete labels. Obviously we may have our own preferences, for instance we may look for maximum area circles, and, among them, for the one with the minimal x-coordinate of the center. But this preference does not affect the equation (*Metaprob1*). Obviously between the set of consistent circles the one which did label the points must exist, but the question "which one?" is ill posed, since up to the observation of the labeled points any consistent circle could have labeled them, and this is again an indeterminacy concern. What concerns compatibility is the measure of the symmetric difference $h \div C$, that is linked to the size of the labeled sample.

If we opt for inconsistency we must control the labeling mistakes.

We may decide gathering the labeled point with another figure, for instance a rectangle,[10] but in this case we are not sure of finding a consistent hypothesis. Rather we must account for labeling mistakes on the part of h and the target of the algorithm may be to output hypotheses with at most k mistakes, paying the additional indeterminacy price as above.

In any case, a rather mechanical operation.

In any case, learning, though a *noble activity* of our mind—in the province of soul, elsewhere—looks like a mechanical interaction between a body with dynamic contour and spikes hindering its motion.

[10] Note that we are inverting the classes with respect to the dressmaker example, for didactic reasons.

3.3.1.4 The Complexity of Learning

To be suitable, the equation (metaprob1) on Sect. 3.3.1.2 must be further elaborated to take into account our unawareness of c, apart from its class $\mathscr{C}$. This requires us to rely on the worst case d (i.e. the number of sentry points for whatever $c \in \mathscr{C}$) which denotes the *complexity* of the class.

Among the notions of complexity of a class of concepts, the most popular is the Vapnick Chervonenkis dimension[11]; the one I prefer is introduced on these pages with the name of *detail* of a class.[12] For short, it's the maximum number of sentry points I need to prevent the expansion of an element of a class into another element of the same class. Until now we have implicitly applied this notion to classes consisting of symmetric differences between either elements of a same class or of two different classes. If the arguments of the difference come from the same class, then the theory establishes that the complexity of the class of the symmetric differences is the same as the one of the one of the original class. This simplifies life for us. However, apart from simple cases, the computation of complexity is a demanding task *per se*, whatever notion of complexity we adopt. Simple examples of details are in the following table

Class complexity depends on how many hooks need to fix the concept contours

Easy to realize	
the detail of the class half lines unlimited on the left is 1	
the detail of a segment of line in 2	
the detail of n segments is $2n$	
Less immediate to realize	
the detail of a circle is 3	
the detail of a k-gon is $\lfloor \frac{4}{3}k \rfloor$	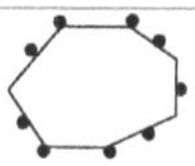

Things worsen if we work with approximate concepts. On the one hand, we remark that the two parameters k and d locally add in (*Metaprob1*). This is the rationale of the *approximate theories* that prove easier to learn when the addition of some mistakes highly reduces the number of sentry points and the complexity of the theory overall. On the other hand, to render the inequality useful we should guarantee in advance the maximum number of mistakes we will be available to bear with our approximate theory, and this may constitute a serious problem.

Approximation may help to learn.

As a matter of fact, results from this theoretical approach to inferring functions from examples, known as *computational learning*, are not directly exploited in application fields, in general. Rather, they help to understand the tools we may put into action to infer functions in actual frameworks, as a function of available data and com-

[11] V. Vapnik, The Nature of Statistical Learning Theory. New York: Springer, 1995.

[12] B. Apolloni and D. Malchiodi, Gaining degrees of freedom in sub-symbolical learning. Theoretical Computer Science, 255:295, 2001.

putational sources, besides the skills of the people involved. Namely, while inferring statistical parameters, such as mean variance or specific parameters of some distribution laws, is a task whose success is related to the availability of the data, i.e. the size of the sample that we may draw on the phenomenon in question, the success of inferring a function—learning it in the current terminology—depends both on the data and on the ability to solve the inverse problem: "Which may be a function, within a given class, that could reproduce, at least approximately, the input-output pairs in the training set?" Said in other words, also in statistical inference the solution in θ of the master equation like the one in λ on Sect. 3.2 is an instance of an inverse problem. But it concerns a limited number of variables: the parameter θ and related statistics. Here, the number of variables may be of the order of the size of the training set. Therefore, the difficulty of solving a learning problem is characterized by three kinds of complexities.

Computational learning: a hard inference;

Namely, we parametrize the quality of the learning task through two indexes

we may appreciate its results in terms of accuracy and confidence, as for results,

- **accuracy**: i.e. a threshold on the probability measure of the symmetric difference between concept and hypothesis,
- **confidence on error**: i.e. the probability that this threshold is satisfied by a compatible function learned by our leaning algorithm

Their values depend on complexities as follows.

Tres sunt genera complexitates

Having fixed to ε and $1-\delta$ the above indexes, we express the difficulty of the learning task as a function of:

and its difficulty in terms of:

1. **sample complexity**, i.e. the size of the labeled sample that guarantee the satisfaction of the above thresholds.

   The *entry level* expression of the sample complexity of a class of Boolean functions is the following

$$m \geq \max\left\{\frac{2}{\varepsilon}\log\frac{1}{\delta}, \frac{5.5(d+t-1)}{\varepsilon}\right\},$$

   It links the sufficient number of labeled examples (hence an upper-bound) to guarantee the satisfaction of the thresholds to: (1) the complexity of the concept class of the target function in terms of detail d introduced above, and (2) the maximum number t of errors of the hypothesis we compute. No any reference is made to the distribution law of the examples; rather it is a distribution free result, hence valid for whatever distribution law, with proper variants linked to the labeling errors (i.e. the label of the examples may be provided incorrectly with some law). Far from being a magic wand, this relation is so universal as to prove unsuitable from a numerical point of view because it results in a too large an upper-bound (say, maybe: "we are sure of meeting the thresholds if we observe a sample of a thousand million examples"). However, the connection between complexity and approximation indexes is meaningful, giving a clear direction to

- size of the training set,

the balance between the detail and the approximation (in terms of allowed errors) of the hypothesis: simply, they add.

2. **computational complexity**, i.e. the computational resources (mainly the runtime of a regular computer) we need to compute a hypothesis, within a class, which is consistent (or approximated up to given number of mistaken points) with the labeled sample. While computing a hypothesis like a half right bounded by a single point is definitely quick and easy, even the task of computing a consistent circle, possibly with further optimality requisites (form instance, in its area) deserves difficulties as for the design of the algorithm and computational costs. The intrinsic difficulty of this kind of problems is established by the NP-completeness of the companion Satisfiability problem for boolean variables.[13] Because of this complexity, it may happen that we prefer relying on an approximate hypothesis also in case of $\mathscr{H} = \mathscr{C}$.

- hardness of coding a solution, and

3. I also put a third ingredient in this trade-off. It consists of the numerical accuracy with which the examples are recorded, say the **numerical complexity**. Let's look at the graph below. It maps the rectangle hypotheses of the dressmaker case study to a two dimensional plane, let's call it the Π chart, where the two coordinates are, respectively, height H and the waistline W of the hypothesis. Since they are the output of our learning algorithm, they inherit the randomness of the sentry points. In greater detail, the red rectangle in the picture is the maximal one consistent with the two sentry points as on Sect. 3.3.1.2; the black rectangles are hypotheses that are consistent with the sentry on the right, but trespass on the sentry at the top. All the corresponding points lie on a line in the Π chart. They exist, but their cumulative probability is null, since the unitary probability mass is spread over the plane through a two-dimensional density function of continuous variables. It is the same reasoning that leads us to conclude that the mass of a point is null inside a continuous one-dimensional body (recall the considerations in point 2 of Sect. 2.3.1.1). Stressing the arguments, we may conclude that a single sentry point is enough to sentinel any concept in the non realistic case where each measure about the examples of the labeled sample have infinite accuracy. In real cases, we work with measuring equipments with finite accuracy and register the data with a limited number of bits, with the consequence that the number of sentry points necessary to meet the thresholds increases up to the *detail* of the class of the target function, with the decrease of the data accuracy.

- accuracy of the available data.

[13] See the introduction to computational complexity in the "further reading" section.

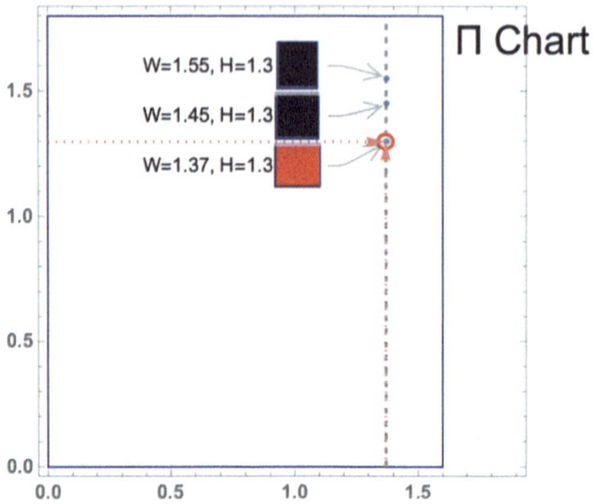

3.3.2 Learning Linear Functions

While many variants of formula (*Metaprob1*) have been considered, both in the literature and in the practice, to take into account:

- conditions making the learning problem less severe, such as special distribution laws of the examples, or special advising actions to avoid mistakes
- condition making the problem more severe, such as malicious mistakes in the labeling of the examples, or availability of solely negative or solely positive examples,

A the root a linear function, like *net*.

the complexity of computing consistent formulas remains a threat that manifests its adverse power here and there. To escape this drawback, computer scientists are used to focusing on the learning of the most elementary function, the linear function

$$y = a + b_1 x_{(1)} + \cdots + b_n x_{(n)},$$

like the one discovered by the kid on Sect. 2.2 (with a conventional shift of symbols a and b), i.e. like the *net* of the neural model. They enrich the problem in two directions: the number of x variables (the independent variables, playing the role of the function input) and proper modification of their space. Note that the attribute *independent* does not refer here to a stochastical independence of the variable; rather it is used to distinguish the input of the function that may be established by us from the output that strictly depends on the input. Moreover, I enclose the variable index into a pair of brackets to indicate that it points at the component of of this variable, thus it has a different role than the one which enumerates the variable specifications.

3.3.2.1 The Discrete Case: Support Vector Machines

A few point to fix the classifier surface:

For Boolean y the problem results in identifying a linear relation between the x components that separates positive from negative points, using the above notation. For scalar x this accounts for a separator point inside a straight line, for two-dimensional x a straight line inside a plane, for three-dimensional x a plane inside a three-dimensional space; in general, a hyperplane inside a multi-dimensional space. The commonly ascertained complexity of the class of hyperplanes equals the number n of the dimension of the space where they are defined. In our approach, where complexity of a class of boolean functions is measured through its *detail*, it is a rather intuitive statement that we may check in the pictures in the following figure as for two-dimensional and three-dimensional space, respectively. In principle a single sentry point is sufficient to prevent a rotation of the hypothesis h that could expand its symmetric difference with the concept c. But we must consider the worst case, where some sentry points may lie exactly on the hinge between h and c. Actually, in the right picture more that two points may lie on the hinge; so, for instance, a fourth point s_4 could do it and we would use 4 points to sentinel this expansion. But we look for a minimal set of sentry points and realize that one point within the set $\{s_1, s_2, s_4\}$ is superfluous. Hence, 3 points are sufficient in the worst case to sentinel a plane (two points to pay forfeit to the hinge and one to block the rotation), 2 points a line and so on.

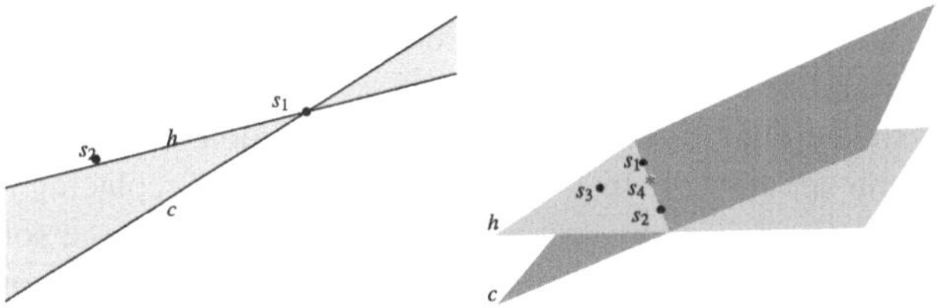

If we compute the hypothesis as the one jointly maximizing the minimal distance from both positive and negative points, the outcoming hyperplane is a function of exactly the points having minimal equal distance from it. So with reference to the right picture in the above figure, we may rely on sentries $\{s_1, s_2, s_3\}$ with the same label to block the shift up of the plane (in fact the plane passing through three points that are non collinear is unique) plus one point of the opposite label to move the plane an equal distance from the two sets of points. While there are many caveats about this argumentation, we may establish from theory that in the general conditions the number of those key points, which we call *support vectors*, is at most $n + 1$. This is why these hyperplanes are the favorite target of learning models, which assume the name of Support Vector Machines (SVM). They denote a moderate sample complexity, which is linear in n. It is still not a disregardable computational complexity, because we must find the support vectors inside the labeled sample, which is smoothed by implementing approximate incremental algorithms.[14]

Small sample complexity, higher computational complexity.

[14] When facing a complex problem, either we are so skilled (and or/so lucky) to find the solution in one shot, or, like suggested by our mom, we do it one step at a time. *Per se*, any sequential algorithm proceeds step by step, but, if we must identify the best 11 players among 20 candidates to

If the separating surface is bent, distort the features, not the algorithm.

But things in our environs are not always linearly separable; their features are more complex. This is a challenge for data analysts and for our basic neuron, the perceptron, as well. The preferred approach of modern data analytics is to keep the discriminator simple, thus preserving the low sample complexity and the suite of incremental algorithms, and to stretch the feature space so as the points arrange themselves in a linear separable space, like below.[15]

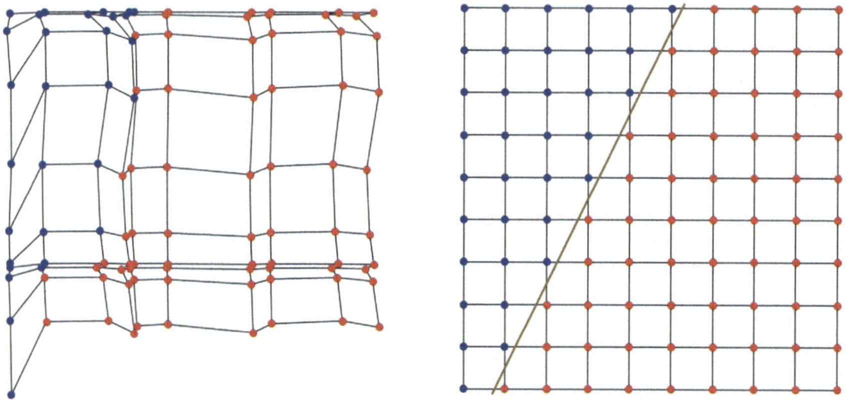

In this way, we move the hard part of the solution upstream, searching for a proper transformation of the feature space that renders the separation problem easy. And since the search entails a difficult inversion problem, again we bypass it by checking the suitability of some general purpose functions for this role. Moreover, since we want to keep the solution of the separation problem easy, a common way passes through the kernel trick. In gross terms, it consists of replacing the *dot product* "·" between pairs of support vectors $x_i \cdot x_j$, at the basis of the solution of the separation problem, with a proper binary function $k(x_i, x_j)$, the *kernel* function, that may be read as a dot product, again, of a transform ϕ of the feature space (the above stretching), namely $\phi(x_i) \cdot \phi(x_j)$. For instance, the Gaussian kernel reads $k(x_i, x_j) = e^{-\gamma d_e(x_i,x_j)^2}$, where d_e is the Euclidean distance between the vectors and γ is a free parameter. Let me remark that, while we preferably work with scalar vari-

---

form the next football team (like for the knapsack problem that will be introduced on Sect. 3.4), the algorithm that sequentially tests the skill level of all possible teams proves too slow. Thus, we look for a procedure that not necessarily leads to the optimal team, but that at each new trial provides a better, or at least not worse, result and produces an acceptable team in short amount of time. The requisite "no worse" could be simply achieved by refusing the new tested group of players if it is worse that the previous ones; the efficiency of the trials, hence the improval rate, depends on *local* selection rules of the players. However, the graph in the last figure of Sect. 1.4 shows the *greedy* rule "substitute one player at a time and confirms the substitution only if the overall skill level of the team increases" may prove faulty, because this level is not the sum of the skills of the individual players; rather, it's the synergistic combination of all the players' skills. We will delve into this aspect when we talk about combinatorial optimization approaches.

[15] Note how the feature space notion formally introduces the concept of representation. The workflow is: we have a phenomenon that we observe according to our propensity. This entails that we observe some features of the phenomenon, which we possibly quantize via metric variables. Both the choice of the features and of variables depend on our convenience.

ables in this chapter so as to remove many mathematical hurdles, here we consider support vectors in a multidimensional space. This is why we need the dot product, whose definition is recalled in the list of symbols, in place of the simple product of numbers.

3.3.2.2 The Continuous Version: Regression

While the SVM as an efficient solution of the classification problem was introduced *only* 30 years ago, the identification of a linear relation between features (the parameters of the perceptron) to explain a continuous variable has been tackled for at least two centuries. Credit for the method of least squares for fitting equations that are linear in the parameters is usually given to C. F. Gauss (around 1809). More in general, regression models are functions that we depute to represent the relation between two equally sized ordered sets of data. We assume their coupling to constitute the pairs of input and output respectively of a function that we know in full apart from a limited set of parameters and a noise affecting the output. The goal of the regression analysis is to infer these parameters from the observed pairs with an accuracy that depends on the noise rate.

In particular, with reference to a scalar independent variable x and the related dependent variable y, the linear regression idea is to draw a line that minimizes the vertical shifts of observed points from it, so that using the $\tilde{y}$ suggested by the line in place of the actual one (either observed or supposed) entails the least damage possible, with a damage/error rooted exactly on these shifts. In the picture below we see this line, described by the equation $y = a + bx$ where a and b are computed as a function of the 20 points (x_i, y_i) observed; the orange arrow links a suggested point $(x_i, \tilde{v}_i)$ to the original companion (x_i, y_i).

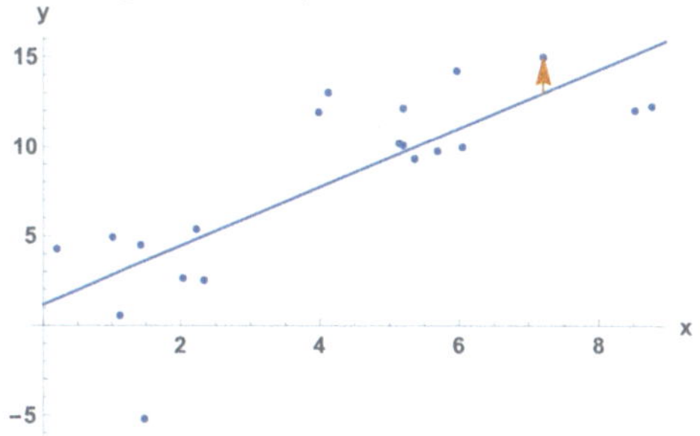

The computation of this function is rather easy and relatively quick; this is true also with less elementary instances, for instance referring to multivariate x. Problems arise when we want to take into account, from an operational perspective, the fact that the parameters of the line are a function of the observed data, so that, if the latter are a random sample, those coefficients are random too. In terms of compatible words, we have a population of compatible lines that may be described through the joint distribution of the compatible (A, B).

Fitting a line is for kids, like on Sect. 2.2,

handling its compatibility is for us.

The approach is the same as in the previous inference tasks: identify a well behaving statistic and derive the parameter distribution by solving the master equations. What change are the continuity of the regressed variables, which removes indeterminacy issues, and the informative richness of the explaining function. Namely, the explaining function of Y now reads either:

$$y_i = g_{a,b,x}(\varepsilon_i) = a + bx_i + \varepsilon_i$$

if we fully know the distribution of ε, or

$$y_i = g_{a,b,x,\theta}(z_i) = a + bx_i + g'_\theta(z_i)$$

if we know the distribution of ε apart from a parameter θ.[16] In both cases g is an explaining function *jointly* reckoning for the *physical* dependency of y_i from the parameters and the *probabilistic* translation of the random seed effects on Y.

The key is proper statistics;

Assuming the ε distribution to be fully known, and x_is as well, by definition, we remain with the pair of parameters (a, b) as representatives of the regression line. Again, learning consists in computing a relevant, possibly well behaving, statistic priming the computation of the distribution law of (A, B). Note that, while in the classification case the statistic coincides with the hypothesis on the concept to be learned, thus resulting in an operational tool for classifying, in the regression case the path to operational tools is a bit longer. For instance, for ε Gaussian with mean $\mu = 0$ and standard deviation σ known,[17]

1. well behaving statistics are

$$s_1 = \sum y_i \text{ and } s_2 = y_i(x_i - \overline{x})$$

   with $\overline{x} = \frac{1}{m}\sum x_i$.
2. A and B, deriving from the related master equations, follow two *independent* Gaussian distribution laws of parameters

$$\mu_A = \frac{\sum_{i=1}^m y_i}{m};\ \sigma_A = \frac{\sigma}{\sqrt{m}} \text{ and } \mu_B = \frac{\displaystyle\sum_{i=1}^m y_i(x_i - \overline{x})}{\displaystyle\sum_{i=1}^m (x_i - \overline{x})^2};\ \sigma_B = \frac{\sigma}{\sqrt{\displaystyle\sum_{i=1}^m (x_i - \overline{x})^2}}.$$

---

[16] For instance $\varepsilon_i = \sigma z_i$, where Z follows a Normal distribution law while σ is the standard deviation of ε.

[17] We recall that a Gaussian random variable is defined by

$$f_X(x) = \frac{1}{\sqrt{2\pi}\sigma} \exp \frac{(x-\mu)^2}{2\sigma^2},$$

where the standard deviation σ of a variable X is the square root of its variance V.

The way of exploiting these distributions is questionable. The (A, B) population in the (a, b) plane, let's call it the π-chart, is like the gray points in the left picture of the figure below. Per usual, we may consider these points as a huge sample from (A, B) distribution law, so large that boiling up frequencies may be confused with probabilities. Now, the confidence of our learning involves peculiar aspects, since probabilities are tossed on the π-chart, meanwhile the operational effects of the confidence are checked on the plane where the regression lines lie, let's call it the r-chart (right part of the figure below). First of all, now confidence intervals revert into confidence regions in both charts. Then, as in previous studies, the problem of identifying a confidence region has infinite solutions, all compliant with the established confidence level, for instance $1 - \delta$. Hence we may opt for different solutions up to our specific targets. Among the others, three distinguishing solutions are shown in the left part of the figure:

the final solution is up to the user.

1. the blue ellipsoid dividing points over a given density probability inside it, from those with lower density outside it. In this way, we radially exclude points with less probability, with the benefit of getting the region with the smallest area having probability $1 - \delta$.
2. the red rectangle that simplifies the above contour, as it depends on 4 edges around the center of the ellipsoid. The distance of the opposite edges can be computed independently, since A and B prove to be independent random variables.
3. the green rhombus with analogous features, as it depends on 4 edges as well, with analogous facility on the distance computation, as the variables $A + B$ and $A - B$ prove to be independent as well.

The preference re the three options may be understood on the r-chart, where the regression line l spans when the parameters span the optioned region of the π-chart, as in the right picture of the figure (here bullets represent the observed points). For the given confidence ($\delta = 0.01$ in the figure), coupling the images of the two pictures through the colors, we see that the ellipsoid region leads to the region in the r-chart where the line may span is most concentrated around the dashed gray line corresponding to the center of the ellipsoid in the π-chart. The green region appears more preferable than the red one, as it hosts lines that are less spread out in the tails.

We remark that there is a bijective relation between points and lines, respectively in the confidence regions drawn in the two charts. In the r-chart we have reported the sampled points as well, and the plain gray line that represents the "original" line. It's the line used in the model above to generate the observation. As we may see, though hidden in all computations, this line is included in the confidence regions, in spite of the small numbers of available observations that are actually compatible with *other worlds* as well. The dashed gray line is the "point estimator" of l that will be introduced in the next section.

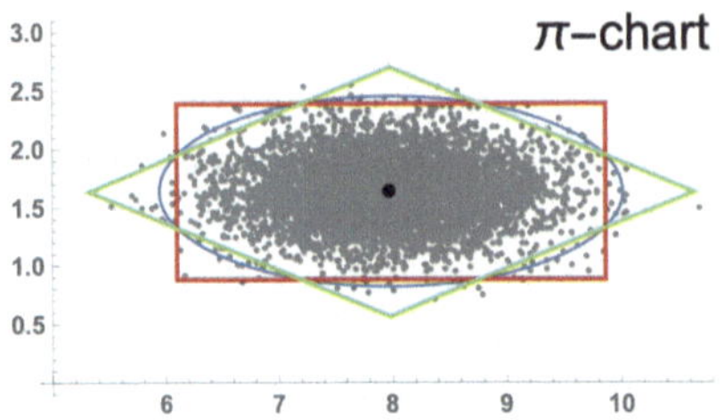

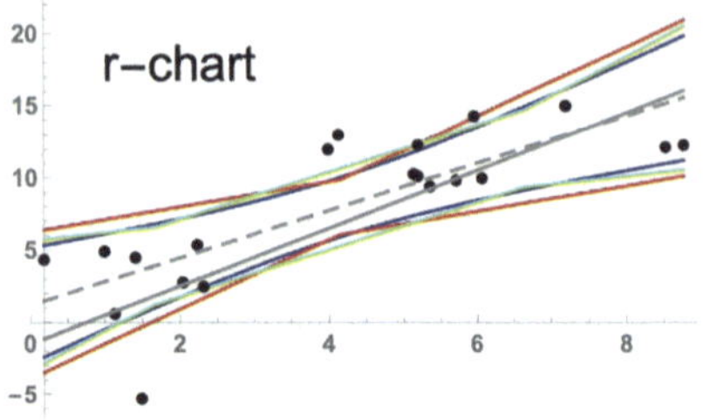

3.3.3 *Steering by Sight*

The concept of *point estimator* first arose formally in the nineteenth century as a tool for replacing set of values generated by complex phenomena with a unique value that produces *almost* the same effect. Needs for such replacements occur in our daily life very often. Thus, aiming to plan the arrival time for a trip from Milano to Bergamo we would skip all local events causing either the braking or the acceleration of our car, we simply want to reason in terms of an "equivalently constant" speed the car may maintain during the trip. The same holds for water consumption per family, in order to decide on the size of a the new aqueduct in a given region, and so on. This is for the concept of estimator of mean. But we may be interested in the maximum battery consumption of our mobile phone so as not to remain without power during the day and in many other kinds of equivalent value.

Looking for an "equivalent value".

From a modeling perspective, we may divide the task into two phases. As for an abstract objective, the target of the point estimator is an overall property of the distribution law of the random variable X which describes the phenomenon we are interested in; this property accounts for a value computed as a function of the distribution law, such as the expected value $E[X]$ or more complex functions. As for an operational objective, we may be interested in a satisfactory equivalent value to be exploited in operational contexts, despite the randomness of the observations we take on the phenomenon, i.e. of the available X sample. Let's denote θ the target value; it represents a parameter of the X distribution, which either coincides with one among those parametrizing the distribution or is a function of them. We denote the equivalent value as the estimator T of θ; it is a function of the X sample, hence a statistic.

We are interested in two aspects:

1. how to compute a T,
2. which one, among the candidate estimators, denotes a specific preferability, thus resulting optimal.

The primordial compressor of randomness is the *sample mean* $\bar{x}$:

$$\bar{x} = \frac{\sum_{i=1}^{m} x_i}{m},$$

i.e. the arithmetical mean: take the sum of observed values and divide by their number. For whatever θ, the vast majorities of estimators is based on this operation. Moreover, thanks to the wide availability of computers, if we are interested in a property of a function of X, we apply this function to the observations x_is and take the mean, as for starter. But science cannot stop at first roadside tavern. Hence the second point is to check for the best estimator of a given parameter θ. Should we pose this question today, the solution would be:

Not only a sample mean;

1. identify the use of this estimator and quantify the deriving benefit,
2. list possible estimators and check numerically which is the winner.

Obviously there may be plenty of objections to this strategy, the former being that we base our selection on the basis of what we have observed, without taking into account what we could observe further. But in the era of simplification this is bypassed by simply enlarging the number of observations (when possible).

Old mathematicians, who where not so breezy and not in the availability of computers, looked for some general optimality criteria to be checked theoretically. All these criteria are pivoted on the idea that the data on which the statistics are based are a random sample drawn from a fixed population to which the unknown parameter refers.

we look for optimal proper-ties, with:

1. The simplest one is the *unbiasedness* of the estimator, i.e. the property that the mean of the estimator T (as a function of the random variables in input) coincides with the parameter θ to be estimated. Accordingly, any sample mean $\bar{x}$ is the unbiased estimator of the mean of a population. Despite, $\frac{1}{\bar{x}}$ is not an unbiased estimator of the parameter λ of the Exponential distribution law in our case study, while $E[X] = \frac{1}{\lambda}$.
2. An analogous criterion is linked to the mean square error of the estimator T of θ, i.e. $E[(T - \theta)^2]$. An optimal estimator in this concern is the one minimizing this quantity.
3. The VIP estimator is linked to the *likelihood* of the sample, where the latter is the probability density of observing the current sample, hence the product if the $f_X(x_i)$ (see Sect. 2.1). The Maximum Likelihood Estimator, denoted as MLE, is a statistic that, replaced to θ in the expression of the likelihood, maximizes the likelihood of the observed sample. It enjoys many properties, including well-behavingness, including that if T is MLE of θ, $h(T)$ is the MLE of $h(\theta)$.

θ fixed, ran-dom sam-ple, or

In our perspective, like in the Copernican *revolution*, the roles are inverted, what is fixed is the sample—hence the estimator t, what is random is the parameter Θ. Thus:

1. Unbiasedness concerns the property $E[\Theta] = t$. Note that $E[\Lambda] = \frac{1}{\bar{x}}$
2. The unbiased estimator t automatically solves the minimization problem

$$\min_{\widehat{\theta}}\{\mathrm{E}[(\widehat{\theta} - \Theta)^2]\}$$

sample fixed, ran-dom Θ.

3. The properties of MLE are clearly explained by its behavior as a well-behaving statistic. Actually, if a parameter θ enjoys a sufficient statistic, this will be a function of an MLE estimator. Thanks to the MLE transfer from θ to $h(\theta)$, we enjoy a sufficiency transfer and a well-behavingness transfer as well.

In conclusion the optimality criteria for estimators—established to get an immediate reward in the agnostic framework—find analytic solutions when related to the distribution laws of the random variables under question, meanwhile they are automatically satisfied by proper functions of the parameters' distribution laws.

Special care has to be devoted to point estimators of functions. As for boolean function, the point estimator we look for is exactly the hypothesis learned by the learning algorithm, while the distribution relates to the classification error, hence a scalar random variable again.

The regression task poses further problems since we want a straight line as a point estimator jointly enjoying good statistical properties and relevant functional meaning. This may be represented by the gray dashed straight line in the "r-chart" of the last figure on Sect. 3.3.2.2, whose coefficients are:

$$\hat{y} = \hat{a} + \hat{b}(x - \bar{x})$$

with

$$\hat{a} = \bar{y}_m = \frac{\sum_{i=1}^{m} y_i}{m}; \quad \hat{b} = \frac{\sum_{i=1}^{n} y_i(x_i - \bar{x})}{\sum_{i=1}^{n}(x_i - \bar{x})^2}$$

Looking for an equivalent function.

The pair $(\hat{a}, \hat{b})$ enjoys two interesting properties:

1. Singularly, they represent the expected values of the distribution laws of A and B, respectively. See previous considerations on unbiased estimators.
2. Jointly, they parametrize the median line of the line distribution law. We may consider it as the pivot of the nested regions we may obtain in the "r-chart" by progressively reducing the confidence level of the confidence regions represented therein.

3.4 From Much Complex to Much Simple, Maybe: *tertium non datur*[18]

Very elementary bricks of our brain,

The discussed linear classification and regression models are in the province of our artificial neuron: a weighted sum in input to a either stepwise of linear activation function. Moreover, they are the root of a plenty of more complex problems, maybe because of the number independent variables, maybe because they are not independent, maybe because the regression function is not linear ... and so on. But

[18] The excluded middle principle which is invoked in many logical contexts.

in this book we are more interested in the root of the problems, rather than in the solution of them. So, conversely, we may look for simpler forms of perceptrons, to delve into their role in surrounding Nature. They exist, are boolean perceptrons, the first model to study our brain introduced by Mac-Culloch and Pitts.[19] In our artificial brain model, neurons sclerotize with aging, thus always produce the same output on the same inputs. If we add some granulation on the communication channels so that only binary inputs and outputs are in the process, we obtain a boolean logic machine. Thus, after many trials, the *expert* mouse in the Y maze experiment on Sect. 2.2, scleortizes its neurons so that, on input "angry yes"/1 (as an alternative to "angry no"/0), of the two possible outputs: (left/0, right/1) they constantly produce the output 0 leading to the cheese. The brains of our early philosophers were more evolute, of course, so that the sclerotized output was an exhaustive function either unary or binary. Like the singer is able to produce many different melodies with the seven musical notes, with the sole functions:

1. unitary negation ($0 \rightarrow 1$; $1 \rightarrow 0$) and
2. binary AND ($(1, 1) \rightarrow 1$, other pairs in input $\rightarrow 0$),

we may construct any function depending on any number of binary inputs and producing a binary output. They may be the suite of functions in the brain of the magistrate producing the verdict guilty/innocent, or the rules of a bank expert system deciding whether to grant a loan customer or not. Perceptrons like in the figure below compute the basic operation NOT, AND and OR (just the NOT of the result of AND applied to the NOTs of the inputs), where the bias of the neuron is represented through a further connection with $w = \theta$ fed by a constant input $c = 1$ and the activation function is Heaviside one (HVS) shown on Sect. 1.3.

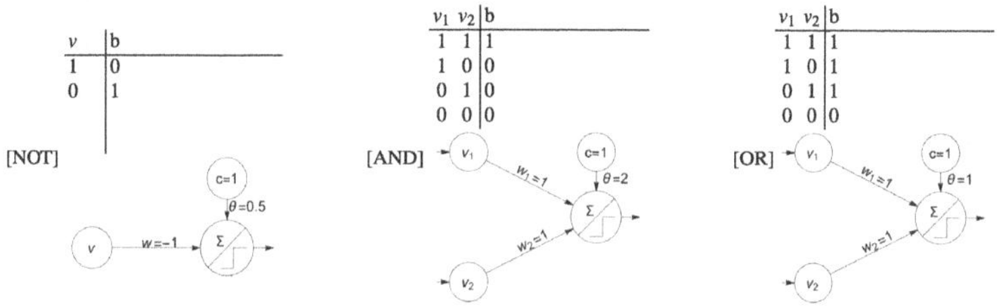

Still better, the two layer perceptron with the structure on Sect. 1.3 may compute any boolean function of its inputs, provided that the output neuron computes an OR of its inputs, and the hidden layer neurons compute the ANDs of their inputs, which in turn devote two nodes (node IDENTITY, node NOT) to each binary variable of the function. This is true because any boolean function may be represented exactly as the OR of the results of a sufficient number of AND nodes, each with a proper subset of connections to the input nodes (Disjunctive Normal Form).

to build any logical argu- ment.

[19] W. McCulloch, W. Pitts, A logical calculus of ideas immanent in nervous activity. Bull. Math. Biophys. **5**, 115–133 (1943).

Who tells us the function of interest to us? With regular neural networks we said that they learn from examples the functions necessary to cope with a given phenomenon (surrounding Nature). With boolean neural networks in principle we cannot learn, apart from elementary networks such as the boolean perceptron, because it's hard to deal with the sensitivity of their parameters to the cost functions.[20] Rather, for a long time the general idea was that these function exist *per se*; indeed they are laws of Nature that we may discover by logical implications starting from a few acknowledged ones (axioms for mathematicians, mysteries for priests) and some agreed inference tools such as the non contradiction law ("tertium non datur"). As a matter of fact, things are more complex, Logics enhanced the minds of mankind to an extraordinary level, by setting up a clean communication language through which cultivated brains exchange ideas and gather a joint work. A tangibly widespread product of this process consists of the computers that currently accompany our lives. Nevertheless, exactly in this framework, the endpoint of the binary logics is represented by the computational complexity.

Logics on the sky,

- "God does not play dice" (if any), according to Einstein.
- Logics do not play chess, since if we want to consider all possible decision trees to adopt the best strategy in response to an adversary we should consider the order 10^{120} move trees. The usual benchmark is the Eddington number that evaluates the number of protons in the Universe to be in the order of 10^{78}.

The knapsack problem is a clever example of problems that are perfectly formalized according to the descriptional logic rules, with no ambiguity or lack of information, yet not solvable in feasible time, i.e. with runtime of a contemporary computer that is less than the Eddington number of seconds under realistic conditions. In more academic terms, it is a np-hard problem.[21] Note that the problem is very common. It's the one met by a delivery company that wants to fill to the maximum the cargo capacity of its truck to get the most profit from a trip. What leaves computer scientists harmless is the fact that, apart from special payloads, the sole way to find the best filling is to tray any arrangement of the parcels into the cargo. For a candidate number of parcels around 300 the number of trials to find the optimum is much higher than the Eddington number.

Thus, the high potentiality of the problem solvable by our computers became the failure point of the wonderful construct raised up by the boolean logic, like a rutilant dance in the Hall of the Arms of a eighty century castle. Obviously, nobody works currently without a computer. Nobody could speak of the decadence of the computer. Maybe, quantum computers will override this failure, at least in part. But computers entered into the play of an over-structure ruled by the cognitive systems. At the end of the day, the warehouse worker filled up the cargo on the basis of his own experience cumulated along the years of his job and shared with the ones of his colleagues.

cognition on the ground.

[20] I said "in principle," since in the literature there are models for training boolean functions, the Boltzmann machine is an archetypal example, but generally they prove less efficient. By converse, the first learning algorithm was deviced by Rosenblatt [23] exactly for a boolean perceptron.

[21] See the introduction to computational complexity in the "further readings" section.

3.5 Which Data?

Up till now, beyond introductory statements, we have subliminally maintained the soul-body dichotomy, talking separately about the conceptual world in terms of information management, realm of the soul, and physical world that our poor body must face. An initial merging of the two realms has been operated, albeit in an abstract thread, by the explaining function of the linear regression model. A distinctly more authoritative one came with the revolutionary relativity theory, by establishing a contraction of the length of a body as a function of a feature of a *noble* activity of its, i.e. its speed in space, thus unifying in a one sole theory the behavior of two quantities which are apparently independent.[22]

A unified framework for physical energy and information,

Long ago, my collaborators and I posed a rather analogous question, concerning how to unify the cognitive behavior with physical behavior of people.[23] We adopted the metaphor of reporters in search of news. Their motion depends on the one hand on where the sources of information are found and, on the other, on the location of the other reporters of the same journal, since they want to avoid redundancy, i.e. do not pursue news that's already been uncovered. The first goal (to get close to the information source) can be studied with a simplistic interpretation of the intentionality attitude of the human brain in terms of a principled symmetry break in a Brownian motion. In the figure below two snapshots show ball set dynamics, respectively with and without symmetry. They correspond to the implementation of a Gaussian increments motion in a square box when each ball is independent of the others, apart from elastic collisions (left part), rather than when some randomly selected pairs of balls become hunter and prey, as in moon park dodgem cars (right part), generating temporary clusters.

to capture intentional motion,

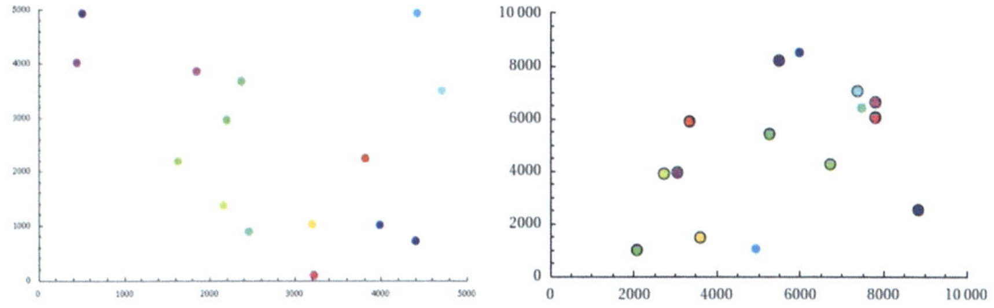

The intertwining with the second goal (to avoid redundancy) is more complex. Thus we decided a downgrading of the scale to shift to a framework that is more familiar to us in this book; namely, we moved from a network of reporters to a network of neurons within a neural network, like the one represented below, where the intentionality is expressed in terms of the mobility as a further attitude of those neurons.

---

[22] Far from wanting to introduce another "lion on the sidelines" topic, I may just suggest this book [28] for a quick introduction.

[23] Please, forget any pretentious mood in the word "unify" on our part.

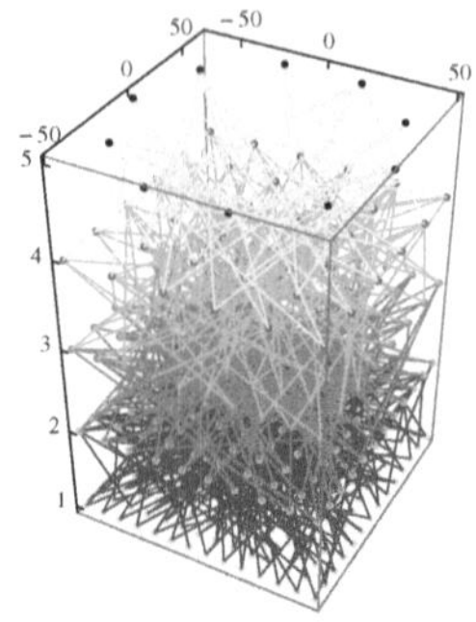

The following figure shows the potential/force field arising in a layer of a multilayer neural network when neurons: (i) are attracted by the upper-layer companions with a force that depends on the value of the information they propagate (gravitational attraction), and (ii) repulse one another as a function of the similarity of their processing (elastic repulsion). This similarity derives from the similarity of the weight vectors associated to their connections to the upper-layer neurons. The deriving dynamics present interesting properties that we may examine as directly referred to the reporters and, more in general, to a social network.

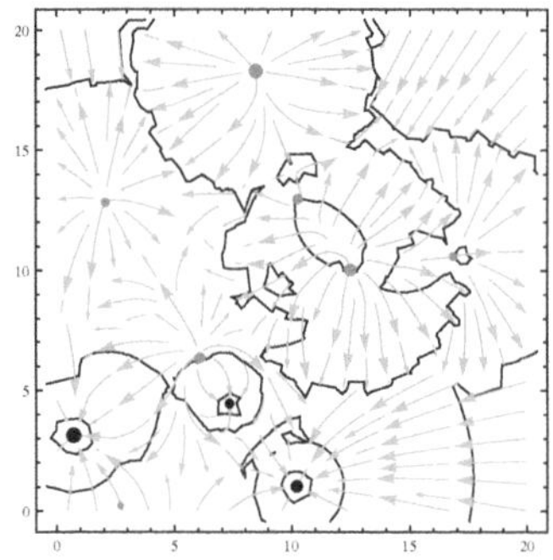

We normally associate the above force fields to the following energy terms P_is that compose the total energy $\mathscr{H}$ of the motion

$$\mathscr{H} = \xi_{p_1} P_1 + \xi_{p_2} P_2 + \xi_{p_3} P_3$$

They respectively account for the potential energy related to a gravitational attraction, the potential energy related to an elastic repulsion and a kinetic energy, where the equilibrium between the two attraction-repulsion forces generated by both attractive upward neurons (black bullets) and repulsive siblings (gray bullets) may be represented through the arrows in the above picture. Here, the bullet size is proportional to the strength of the forces, hence either to the neuron mass (black neurons) or to the outgoing connection weight averaged similarity (gray neurons). Black contour lines are equipotential curves.

However, here the masses and elastic strengths are of a cognitive nature, hence they are linked to the value of the piped information and the computational parameters of the single agents (either neurons or reporters). The velocities in P_3 are physical quantities, which present either precisely the neuron velocities, in the original model, or the virtual approaching to an interest community via terminals in the social network. To these quantities a fourth form of energy E_c adds which is represented by the cost function considered on Sect. 1.4.2.2. Given the non conventional variables involved, we mold the sum through suitable constants ξ.s. In this way we consider a unified adaptive process consisting of motion and learning, in search of the minimum total energy

where all kinds of energies merge.

$$\mathcal{H} = \xi_{e_c} E_c + \xi_{p_1} P_1 + \xi_{p_2} P_2 + \xi_{p_3} P_3$$

where the link between the information process controlled by E_c and the dynamical process ruled by the P_is is expressed by the above intertwined parametrization and by the Elementary Processing function (EPF see Sect. 1.3) of the involved neurons. The new EPF reads

$$y = \sigma\left(\sum_{i=1}^{n} x_i w_i \lambda_i + b\right)$$

where λ_i is a coefficient penalizing the the distance between the mobile neuron i and its upper-layer companion attracting it. It might look like an extended version of the Hamiltonian of the system where the functional aspects (analysis of the functions satisfying $\mathcal{H}$) are replaced by cognitive ones (learning of the trajectory in the extended physical-knowledge space).

The ambition of this modeling is to capture the dynamics of social communities, such as the abovementioned community of reporters or social networks in general, where the individual members sniff on the WEB the pages in sync with their interests so as to join the respective groups. At the same time, people look for novelty, hence are not attracted by other people in a group who process information exactly as they do. In this perspective, the snapshot in the figure below may well reflect the inner dynamics of communities. Actually, the picture refers to two locations of the neurons in the input layer of a multilayer perceptron trained to identify the value of the digits handwritten on a square of 14 × 14 gray level pixels. We have two kinds of dynamics: (i) the one of particles/neurons obeying to a Newtonian-like mechanic given the involved positions and moments, and (ii) the one of parameters of the system, typically the connection weight matrix and the threshold vector of the neural network which change to achieve a given learning task. Both are ruled by the same Hamiltonian, in turn evolving in search of its ground state as the minimum of the extended Hamiltonian. In the Output layer the digits are associated to 10 neurons distributed along a circle. We may see how the input neurons move from a regular grid (in the left picture) to clusters located in correspondence to 6 of the 10 digits (in the right picture). The latter are denoted by red balls, the former are tracked by colored trajectories.

Social community dynamics

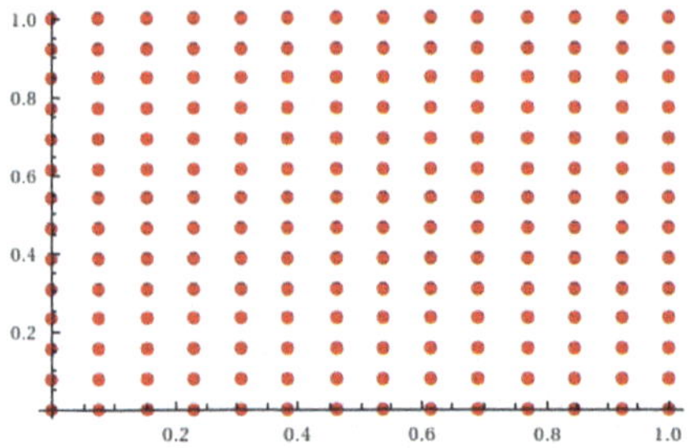

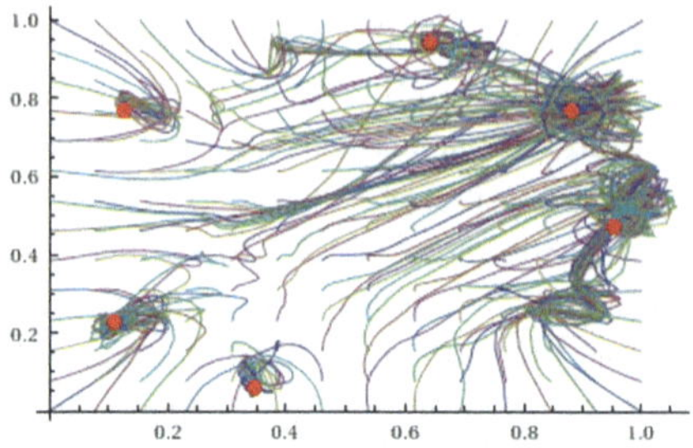

The aim of the neurons to differentiate their role results in not clustering themselves under a single digit. The effect of this network strategy is a good performance in recognizing the ten digits of the well-known MNIST benchmark[24] in a small number of learning cycles.

Remark 3.1 Note how the neural dynamics plays a role similar to the *environment conditioning* of *glia* cells in the brain mentioned on Sect. 1.2. Namely, our artificial quad-partite synapses gather the convergence of the conventional signaling of the wired neurons plus additional backward signals deriving from the dynamic components of the extended Hamiltonian and the diffuse effects of the neuron topology inside a same layer deriving from the new EPF above.

3.6 Further Readings

3.6.1 The Chicken Egg Dilemma

In this chapter we remain in the conventional realm of data ruled by probability distributions, albeit with a different approach. The picture below sketches the difference between a conventional approach and ours.

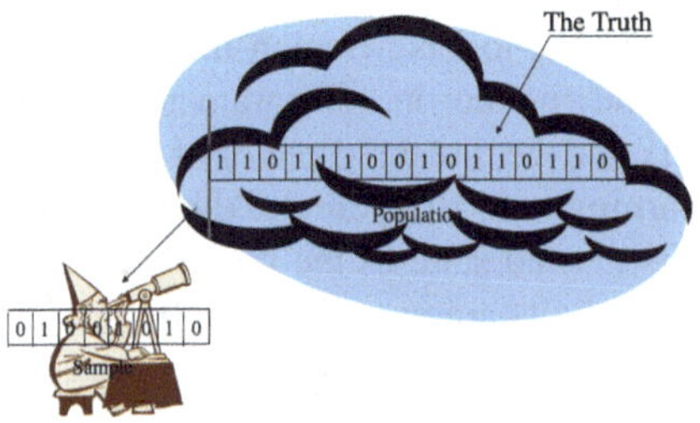

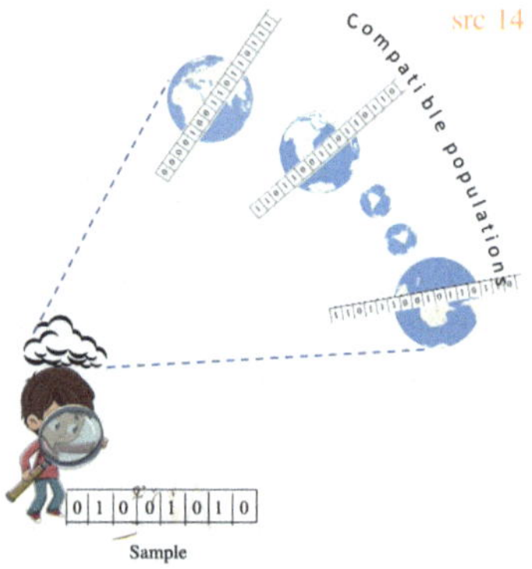

---

[24] http://yann.lecun.com/exdb/mnist/.

In very essential terms, we go from a mental scheme where after first establishing all the rules, God tosses dice for giving us feeble insights (left part of the picture), to an operational framework where rules are random functions that we have tools to manage in order to hit our own target constituting the temporary *unquestionably place of the universe*[25] (right part of the picture). Thus,

Inverting the perspective in search of explanations.

Population first. Conventional statisticians assume the sample to be an excerpt of the random population existing in Nature (at least in the clouds), and pay any effort to derive from these available observations the true shape of the population distribution law—for instance the parameter θ of the related CDF $F_\theta(x)$, provided F is known. *(fixed parameters—random data)*

Data first. Computational-learning-drawn approaches consider the sample to be all that's available, so we look for a suitable synthesis of them in terms of compatible distribution laws. If we assume we have accumulated enough observations to be highly confident about the shape of the distribution, our inference concerns the parameters of the distribution which may assume different values that are compatible with the values of the data we have observed. *(fixed data—random parameters)*

Per se, contrasting the two approaches may appear to be a chicken egg dilemma as futile as the Lilliputian contest described by Jonathan Swift in the "Gulliver travels" about the way of breaking eggs: upon the larger or smaller end. In place of the six rebellions set off by Lilliputians over this issue, we simply claim that the second approach may improve the value of the estimators. For instance, it complies with contemporary customs and ideas in an age when people are bombarded by data, in the form of market advices for instance, and their main goal is to understand who is the sender, possibly his reliability, the real convenience of his offers, and so on—in one word, the whole algorithm generating these data. Since the *true* profile of the sender will never be achieved (note: not only because it is hidden by remote connections), we mind about him in terms of compatible profiles and take decisions accordingly. This may be the essence of *learning* the profile, which we tackle by identifying in a probabilistic way a model explaining the bombarding data.

3.6.2 Delving into the Roots

Wanting to start from the very beginning to illustrate our perspective, we assume a set of observations to be the sole knowledge we have about a possible experiment. The formalization of them as a realization $\{x_1, \ldots, x_m\}$ of a sample of random variables $\{X_1, \ldots, X_m\}$ means in practice that we give each observation the same weight.

[25] It is unquestionably the place of the universe, to which the motion and position of all the other heavenly bodies are compared... [N. Copernicus, De Revolutionibus Orbium Coelestium, Book 1, Chapter X,1543].This incipit of the book preface to Pope Paul III sparked a long and often cruel ideological debate on the part of people used to replacing the force of the science with the force of the power—per usual [27].

Weights, in turn, are the operational counterpart of the probabilities with which we characterize the data in a non deterministic framework. In particular, the equal-weight condition is commonly synthesized by the empirical cumulative distribution function $\widehat{F}_X$, introduced on Sect. 3.1.2 as:

Starting from a sample,

$$\hat{F}_X(x) = \frac{\text{number of } x_i s \text{ such that } x_i \leq x}{\text{number of } x_i s}$$

It corresponds graphically to the black curves in the figure below [1] computed on a sample of 30 observations.

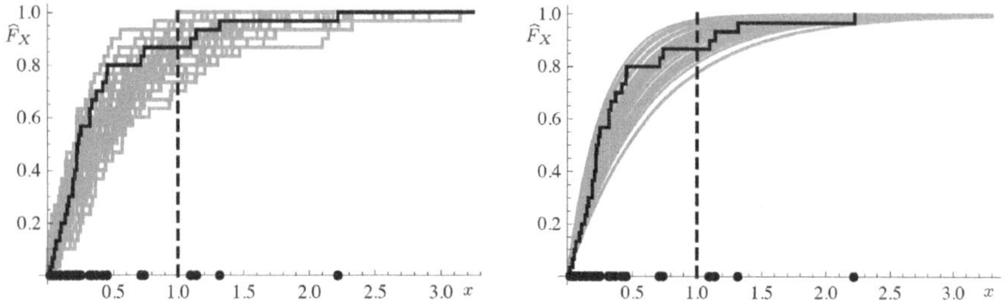

what we would observe from further samples,

We are aware that, since we are considering random variable realizations, we could face values different from those actually observed. For instance, if the X_is follow a negative exponential distribution with parameter $\lambda = 2.0$, by plotting an analogous ECDF for 40 more samples observed in the same experiment, we obtain curves like the gray ones shown in the left part of the figure. All these curves are characterized by steps having the same height equal to $1/30$ (since $m = 30$) and different widths, as they result from the different locations of the data on the horizontal axis. The tool that is commonly proposed to deal with this family of curves, hence with the variability of the samples in spite of the uniqueness of their originating population, is the confidence interval. In principle, it consists of an exchange of the variability on the horizontal axis contrasted with the fixed values on the vertical axis, leading to fixed values on the former and variable ones on the latter. In this way, we may say that, according to what we have seen, the probability π of having X less than a given x (say 1.0) ranges between 0.73 and 0.97, as may emerge from the intercept reckoning of the gray curves with the vertical line $x = 1.0$.

rather, what we observe from the *actual* sample.

In the right part of the Figure we propose another way of managing the uncertainty. Both graphs refer to cumulative distribution functions (CDF), yet in two different statistical contexts. With the former graph we are in the situation where we know the true λ, draw a series of samples and compute an ECDF from each of them. In a more realistic scenario, we do not know λ (indeed, we want to estimate it). Rather, we know exactly what we have observed and ask ourselves which random variable—hence which CDF—could have produced these observations, provided we also know that its distribution belongs to the negative exponential family. The gray curves in the picture are the answer we get by using compatible values of λ. They entail a different interval for π.

It is a way somewhat different from the classical one. Nevertheless, it benefits from being easier to understand as well as to implement, even without the usual assumptions of *quasi* Gaussianity of the involved statistics distribution law [16]. Indeed, inferring confidence intervals is not always an easy task within classical frameworks. Its solution was the focus of a major dispute in the mid 20th century about the interpretation of parameter variability [30] in terms of *fiducial* distribution [10], *structural* probabilities [12], *priors/posteriors* [22], and so on. From an epistemological viewpoint, this entailed a companion dispute as to the nature of probability: Is it a physical feature of the phenomena to be described through random variables or a way of synthesizing data about a phenomenon? For a long time, Neyman's confidence intervals won out for all practical purposes, crediting the phenomenological nature of probability. As a matter of fact, his approach is best suited to dealing with tests of hypotheses—producing for instance the keystone Neyman–Pearson lemma [20]—and has a few results about confidence intervals as corollaries derived from a dual perspective on the same parameter identification problem [17]. With a fixed but unknown parameter and known but random data we may: (1) check whether the observed data are compliant with the hypothesized value of the parameter—direct perspective; or (2) stretch the value of the parameter until the observed data maintain acceptable compliance with it—reversed perspective. The results we obtain with the second perspective are very limited, since they apply only with (possibly broad) approximations when the candidate models for explaining data cannot involve Gaussian probability distributions or tail bounds [16]. For instance, we may note that most results of computational learning theory are based on the Chernoff/Hoeffding inequality [6] simply as a sophisticated variant of the Chebyshev inequality. Bayes' approach offers a valid escape from this drawback by considering a priori parameter distribution laws—sometimes linked to the features of the subtending phenomena [3, 11], other times supported either by clever entropic arguments (the *uninformative* priors [14]) or by tight analytical considerations (the *conjugate* priors [8]))—which are modified by the information content of the observed data [7]. Nevertheless, very often this solution just ends up shifting the hard part of the inference problem one step back, i.e. to the answer to the question: Who tells us the priors?

Two long dated perspectives,

In our approach we essentially propose to return to Fisher's original idea of considering the model parameters to be random variables *in light of the observed data* [10]. From this perspective, he defines a fiducial distribution law of parameters of a given random variable that he deduces from a sample of its realizations. With this law he computes, for instance, "the probability that μ (mean of a Gaussian variable) is less than any assigned value, or the probability that it lies between any assigned values, or, in short, its probability distribution, in the light of the sample observed" [9]. This perspective contrasts with the above scheme of fixed but unknown parameters. Rather, it is in line with the Bayesian approach, whereas the main difference in its respect is that we do not need any kind of prior to compute the parameter distribution. However, the prior plays a mediating role in the mentioned dispute on the nature of probability, since it constitutes a pre-existing source of parameter randomness, whose final distribution may be modified again in light of the observed data. By contrast, we derive this distribution uniquely from suitable statistics such as the sufficient statistics [29]. Actually, the major Fisher–Neyman fight [4, 19] on the matter may appear no different *per se* from the mentioned Lilliputian contest. Here again, we merely claim that, of the related compatible worlds, we opt for the Fisher-like one (our proposed world) for two reasons:

with different operational consequences.

- a meta-reason: it is a world routed on compatible worlds
- it provides an at once *easy* and rigorous rationale to the computational learning in the core of the modern scientific societies.

Now, we can.

I remark that our approach is routed on the solution of inverse problems. This entails complexity issues that are bearable, however by the current computing facilities.

3.6.3 A Few Basic Statistics

As told earlier, any output of a code having in input sampled data may be considered as a statistic on that sample. On the other hand, like with the basic computations (aka the perceptron, or the unary and binary logical gates) whose composition gives rise to the variegate universe of all computations, most of the statistics in use are the composition of three elementary ones:

1. sample sum: $\sum h(x_i)$, for any function h,
2. sample maximum: $\max\{x_1, \ldots, x_m\}$ and
3. sample minimum: $\min\{x_1, \ldots, x_m\}$.

In many operational conditions, they prove to be sufficient statistics. This property is supported by the following theorems:

Standard sufficient statistics

Lemma 3.1 *For any family of probability distributions whose density functions can read as*[26]

$$f(x;\theta) = \exp\left(c(\theta)d(x) + a(x) + b(\theta)\right)$$

where a, b, c and d are real-valued functions, the statistic $\sum_{i=1}^{m} d(x_i)$ *is a sufficient statistic for* Θ.

Lemma 3.2 ([15]) *Given a sample* $\{x_1, \ldots, x_m\}$ *with explaining function* g_θ *such that X ranges in the interval* $[a, \theta]$ *for* $a \in \mathbb{R}$, *we can compute a sufficient statistic from it if and only if a factorization of the density of X exists such that, for any truncation parameter* $b \in [a, \theta]$

$$\frac{f_{X_\theta}(x)}{F_{X_\theta}(b)} \text{ is independent of } \theta$$

Standard sufficient statistics, again

If this statement is true, then a sufficient statistic for Θ *is a function of* $x_{(m)} =$

[26] They define the *Exponential Family* of distributions.

$\max_{i=1,\ldots,m}\{x_i\}$.

Analogously for X ranging in the interval $[\theta, b]$ *for fixed* $b \in \mathbb{R}$, *if, for any truncation parameter* $a \in [\theta, b]$

$$\frac{f_{X_\theta}(x)}{1 - F_{X_\theta}(a)} \text{ is independent of } \theta$$

then a sufficient statistic for Θ *is a function of* $x_{(1)} = \min_{i=1,\ldots,m}(x_i)$.

Thanks to the transfer properties mentioned on Sect. 3.3.3 we may generally expect that the composition of these basic statistics might give rise to further sufficient statistics.

3.6.4 A Few Suitable Kernels

Many kernels, in principle; a few ones, in practice.

The general suitability of a kernel with our scope comes from its mentioned interpretation in terms of a dot product of a function ϕ applied to points y_i and y_j, which projects y into a higher dimensional space (called feature space). Typical kernels of wide use, beyond the above mentioned dot product, are listed in the following table.

Inhomogeneous polynomial kernels of degree d	$k_p(y_i, y_j) = (y_i \cdot y_j + c)^d$
Gaussian kernel	$k_G(y_i, y_j) = e^{-\beta l^2(y_i - y_j)}$
Sigmoid kernel	$k_s(y_i, y_j) = \tanh(\alpha y_i \cdot y_j + \beta)$

- Inhomogeneous polynomial kernel of degree d may be regarded as a dot product of the $\binom{n+d}{d}$-dimensional space of all monomials in n variables (where n is the dimensionality of the original data), up to degree d with c arbitrary constant.

On the contrary, the feature space associated to both Gaussian and sigmoid kernels are of infinite dimensionality and may be reformulated through quite complex $\phi(y)$. In particular

- the shape parameter h in Gaussian[27] kernel relates to the width of radial basis-like functions, while
- scale and position parameters α and β of sigmoid kernel allow for a stretching/translation of the sigmoid function at the basis of neural networks. Note that, despite its wide use, the sigmoid kernel does not belong to the class of positive semidefinite bilinear functions, thus losing most properties strict-sense kernels hold.

[27] Where, for $z = (z_1, \ldots, z_n)$, $l^2(z) = \sum_{i=1}^n z_i^2$.

The reader may find a complete definition of kernels in [26].

Kernels are widely used in the machine learning community since, given an algorithm formulated in terms of dot products, we can always construct an alternative algorithm by replacing the dot product $\cdot$ with a kernel k (*kernel trick*) [25]. This proves very efficient for extending in quite a natural way and without extra efforts the strict functionalities of a linear algorithm to handle nonlinear problems.

3.6.5 Computational Complexity

In this appendix we just outline a few definitions that let us formally identify the computational complexity of a numerical task with sole reference to the necessary computation time. Thus we will not mention the other important resource binding our numerical implementations represented by the mass memory. Computational complexity is a broad research and application field. For starters, the reader can take a look at two classical books cited in the Refs. [13, 21]. The definitions we give refer to any common general purpose computing device like the PC on our desk. We give a formal description of its properties by referring to a very essential yet equivalent device: the Deterministic Turing Machine. Since it's a very standard matter, differently from the rest of book, here we prefer to maintain the itemization of definitions and pictures as drawn from another book of mine [2].

Definition 3.1 A *Deterministic Turing Machine* is a computing machine that consists in a finite state control device (mind the CPU of your PC), plus a read-write head operating on an infinite tape divided into spares indexed by integers (your large—though non infinite—capacity hard disk) (see Fig. 3.1).

Definition 3.2 A *program* $\mathscr{A}$ for a DTM, generally speaking an algorithm, is given by the following specifications:

1. A finite set T of tape symbols including a blank.
2. A finite set S of states, including a start-state s_0 and two halt states: yes s_y and no-state s_n.

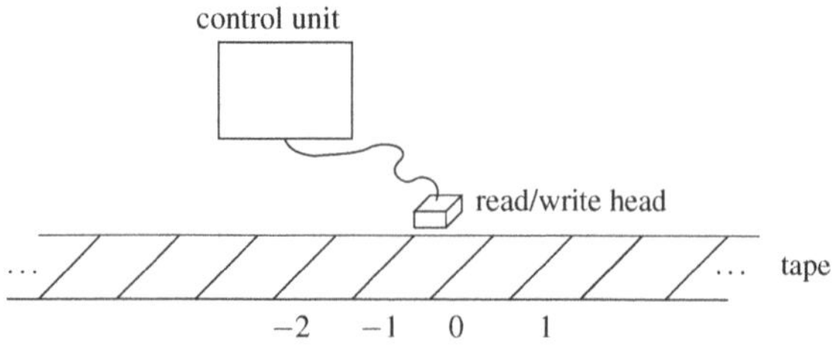

Fig. 3.1 Scheme of the Turing Machine

3. A transition function $\delta : (S - \{s_y, s_n\}) \times T \mapsto S \times T \times \{-1, 0, +1\}$. If we denote by T^* the set of all finite strings from T, including the empty string, the *input* (otherwise named *instance* of $\mathscr{A}$) is a string $x \in T^*$ whose first element is in the square 1 of the tape and the rest are consecutively stored on the right of it; all the squares either preceding or following x contain the blank symbol.

A DTM computation runs in the following way. $\mathscr{A}$ starts with the control device in the state s_0, and the read-write head scanning tape at square 1. The control device evolves according to δ, which on the basis of the current tape symbol and state determines a new state, the symbol to be printed and the successive shift of the head to the left (-1), to the right ($+1$) or no move (0). Each δ application represents one step in DTM. $\mathscr{A}$ with input x—which we denote $\mathscr{A}(x)$—stops when the control device enters the state s_y or s_n, which respectively denote the string x as *accepted* or *rejected* by $\mathscr{A}$. In this sense we say that $\mathscr{A}$ solves a *decision problem*.

We collect the set of instances of a computational problem into a *language*. Namely

Definition 3.3 For any finite set T of symbols, a *language* L is any subset of T^*. If each element of L is accepted by $\mathscr{A}$, then we call the language $L_{\mathscr{A}}$ and say that it is *recognized* by $\mathscr{A}$.

Definition 3.4 For a language L, and its related decision problem, the (time) *computational complexity* of L is a function t from $\mathbb{N}$ to $\mathbb{N}$ such that:

$$t(n) = \min_{\mathscr{A}\ recognizing L} \left\{ \max_{x \in T^n \cap L} \{\text{number of steps before } \mathscr{A}(x) \text{ halts}\} \right\} \tag{3.1}$$

For a DTM program $\widetilde{\mathscr{A}}$ which stops for each input $x \in T^*$, its (time) complexity $\tau(n)$ is defined as follows:

$$\tau(n) = \max_{x \in T^n} \{\text{number of steps before } \mathscr{A}(x) \text{ halts}\} \tag{3.2}$$

We distinguish between easy and difficult languages (and then problems) in function of their computational complexity.

Definition 3.5 The class P is the class of decision problems L such that there exists a polynomial complexity program $\widetilde{\mathscr{A}}$ recognizing L. For short, you accept or reject with $\widetilde{\mathscr{A}}$ any $x \in T^n$ in a time polynomial in n.

We essentially identify with P and its search version the class of feasible problems.

Definition 3.6 A *search problem* is a concatenation of decision problems, each generating an additional bit of the solution. A search problem has a polynomial complexity if the length of its solution is polynomial and the complexities of the involved decision problems are polynomial too.

Besides these classes we have a wide spectrum of unfeasible problems whose complexity ranges from a superpolynomial function up to any kind of unbearably growing functions. A very common family of problems, including problems that on the one hand occur in everyday life and on the other are open to getting approximate solutions in feasible time is rooted in the NP class. It is a class of decision problems whose complexity is appreciated in the ideal computational framework represented by a non deterministic Turing machine, i.e. a Turing machine that performs many computations in parallel. This attitude is described by the following definition.

Definition 3.7 A *Non Deterministic Turing Machine* (NDTM) is a DTM endowed with more than one read-write head and a parallel transition function $\delta : \{S - \{s_y, s_n\}\} \times T \rightarrow 2^{S \times T \times \{-1,0,+1\}}$.

Thus, with a tape character in input and a non terminal state, the function might generate more than one triplet (state, character to print, next move). This means that the read-write head splits in many replicas which print characters and move along the tape.

On this machinery the NP class is an extension of the P class so defined:

Definition 3.8 Class NP is the class of decision problems L having polynomial complexity on an NDTM, i.e such that for each L there exists a NDTM program $\mathscr{A}$ accepting each $x \in L$ in a time polynomial in the length of the x.

None showed that class P does not coincide with class NP (the famous P $\neq$ NP dilemma) but neither did anyone prove the contrary. Namely:

- a special set of problems, the NP-complete class, have the property that any other problem in NP can be quickly (that is, in polynomial time) converted into each one of them. Thus if an algorithm exists solving an NP-complete problem in polynomial time, then other algorithms must exist solving quickly any problem in NP
- no one has ever been able over the course of about half a century (since the formalization of the class) to discover an algorithm solving an NP-complete problem in polynomial time.

So we are used to conjecturing that such an efficient algorithm does not exist, hence P $\neq$ NP.

The extension of the NP class to search problems gives rise to two (possibly coinciding) classes of problems.

Definition 3.9 A problem belongs to the NP-easy class if its solution requires the solution of a polynomial number of NP decision problems.

Definition 3.10 A problem belongs to the NP-hard class if the existence of any algorithm solving it in polynomial time would imply the existence of an algorithm solving a NP-complete problem in polynomial time too, hence P $=$ NP.

Remark 3.2 In spite of the names, an NP-easy problem is NP-hard too.

A typical way to overcome the complexity of NP problems is to introduce some random steps in the computation. While not ensuring any solution in feasible time, it offers in many cases a good probability of the same. This is due solely to the fact that random steps break any possible stubbornly wrong calculation strategy. The extension of a DTM to a probabilistic Turing machine is the following.

Definition 3.11 A *Probabilistic Turing Machine* (PTM) is a Turing machine whose transition function is: $\delta : (S - \{s_y, s_n\}) \times T \times R \to S \times T \times \{-1, 0, +1\}$ where R is an infinite set of independent random bits having equal probability of being set either to 0 or 1.

For each step, δ takes in input one of these bits in addition to the current character and state for computing the triplet (next state, next character, next move). Actually, δ may use or ignore the drawn bit, depending on the algorithm implemented.

There is a very detailed taxonomy of the classes of problems solvable by this machinery, depending on the running time and correct solution probabilities. Here we just mention the RP class that is recalled in the book.

Definition 3.12 A problem π belongs to the RP class if an algorithm $\widetilde{\mathscr{A}}$ exists running in polynomial time on a PTM such that denoting by L_π the related language, $\mathrm{P}(\mathscr{A}(x)$ outputs $s_y | x \in L_\pi) \geq \frac{1}{2}$ and $\mathrm{P}(\mathscr{A}(x)$ outputs $s_y | x \notin L_\pi) = 0$.

DTM, *function* and *algorithm* are three different ways of meaning a univocal map from input to output of our computer. DTM mostly evokes the physical mechanism of the computation. Rather, function defines the mapping in mathematical terms; while algorithm describes it in terms of logical instructions. Consider the set of all function computable by a DTM. According to the Church–Turing thesis [5] they represent all functions computable by any computing device. They are denoted as *partial recursive functions* (PRF) as they: (i) result from the iterative application of a limited set of primitive functions—recursiveness—and (ii) their output is not necessarily defined for each input (which happens whenever your PC enters a neverending loop)—partiality. Now your PC may compute any one of these functions, provided you fed it with a higher level input constituted by the *computation program*. Namely, before supplying the input of the function you want to compute, you supply the machine with a set of instructions for simulating the DTM that computes exactly this function. For this reason you speak of your PC and any DTM with these properties as a *universal DTM*. Namely:

Definition 3.13 A *universal DTM* is a DTM able to compute any PRF, i.e. to simulate any other DTM.

Obvious extensions exist of this universality notion to *universal NDTM* and *universal PTM*. In particular our PC is a *universal Deterministic Turing Machine*, according to Definition 3.1, apart from the length of the tape.

References

1. B. Apolloni, S. Bassis, Confidence about possible explanations. IEEE Trans. Syst. Man Cybern. Part B-Cybern. 1–15 (2011)
2. B. Apolloni, D. Malchiodi, S. Gaito, *Algorithmic Inference in Machine Learning*, vol. 5 (Advanced Knowledge International Pty, Adelaide, 2006)
3. J.O. Berger, *Statistical Decision Theory and Bayesian Analysis*, 2nd edn. (Springer, New York, 1999)
4. J.F. Box, R.A. Fisher, *The Life of a Scientist* (Wiley, New York, 1978)
5. W.S. Brainerd, L.H. Landweber, *Theory of Computation* (Wiley, New York, 1974)
6. H. Chernoff, A measure of asymptotic efficiency for tests of a hypothesis based on the sum of observations. Ann. Math. Stat. **23**, 493–509 (1952)
7. F. Cuzzolin, Two new Bayesian approximations of belief functions based on convex geometry. IEEE Trans. Syst. Man Cybern. Part B: Cybern. **37**(4), 993–1008 (2007)
8. P. Diaconis, et al., Conjugate priors for exponential families (1979)
9. M.A. Fisher, The fiducial argument in statistical inference. Ann. Eugen. **6**, 391–398 (1935)
10. R.A. Fisher, Mathematics of a lady tasting tea, in *The World of Mathematics*, eds. by J.R. Newman (Simone & Schuster, New York, 1956), pp. 1512–1521. Original work published in 1935
11. J.P. Florens, M.M. Mouchart, J.M. Rolin, *Elements of Bayesian Statistics* (Marcel Dekker Inc, New York, 1990)
12. D.A.S. Fraser, Structural probability and a generalization. Biometrika **53**(1/2), 1–9 (1966)
13. M.R. Garey, D.S. Johnson, *Computer and Intractability: A Guide to the Theory of* NP-*Completeness* (W. H. Freeman, San Francisco, 1978)
14. E.T. Jaynes, Prior probabilities. IEEE Trans. Syst. Sci. Cybern. **4**(3), 227–241 (1968)
15. M.G. Kendall, A. Stuart, *The Advanced Theory of Statistics*, Volume 2: Inference and Relationship (Charles Griffin, London, 1961)
16. W.Q. Meeker, L.A. Escobar, Teaching about approximate confidence regions based on maximum likelihood estimation. Amer. Stat. **49**(1), 48–53 (1995)
17. J. Neyman, Su un teorema concernente le cosiddette statistiche sufficienti. Giornale dell'Istituto Italiano degli Attuari **6**, 320–324 (1935)
18. J. Neyman, Outline of a theory of statistical estimation based on the classical theory of probability. Philos. Trans. R. Soc. Lond. Ser. A, Math. Phys. Sci. **236**(767), 333–380 (1937)
19. J. Neyman, Fiducial argument and the theory of confidence intervals. Biometrika **32**(2), 128–150 (1941)
20. J. Neyman, E.S. Pearson, On the problem of the most efficient tests of statistical hypotheses. Trans. R. Soc. Lond. Ser. A **231**, 289–337 (1933)
21. C.H. Papadimitriou, *Computational Complexity* (Addison-Wesley, Reading, 1994)
22. F.P. Ramsey, The foundations of mathematics, in *Proceedings of London Mathematical Society* (1925)
23. F. Rosenblatt, The perceptron - a perceiving and recognizing automaton. Technical Report 85-460-1, Cornell Aeronautical Laboratory, Ithaca, New York (1957)
24. M. Rosenblatt, Remarks on a multivariate transformation. Ann. Math. Stat. **3**(23), 470–472 (1952)
25. B. Scholkopf, A. Smola, *Learning with kernels* (MIT Press, Massachusetts, 2003)
26. B. Scholkopf, A.J. Smola, A short introduction to learning with kernels, in *Machine Learning Summer School*, eds. by S. Mendelson, A.J. Smola. Volume 2600 of *Lecture Notes in Computer Science* (Springer, Berlin, 2002), pp. 41–64
27. J. Stanley, *How Fascism Works: The Politics of Us and Them* (Random House, New York, 2018)
28. R. Stannard, *Relativity: A Very Short Introduction* (Oxford University Press, Oxford, 2013)
29. S.S. Wilks, *Mathematical Statistics*. Wiley Publications in Statistics (Wiley, New York, 1962)
30. S.L. Zabell, R.A. Fisher, Fiducial argument. Stat. Sci. **7**(3), 369–387 (1992)

Chapter 4
How to Tackle Difficult to Understand Phenomena. Fuzziness, Cognitivity, Memory

Abstract In worlds that are compatible with the assumption that we have overall visibility of it and the m data we observe appear without any constriction, so that they have the same relevance $1/m$, probabilistic models are the landlords. Rather, if we have a limited scope we must ask other people to add data within their scope, and the relevance of the overall observations is no longer uniform. In this way we enter the modern connected worlds, made up of social networks and fuzzy graduations. In this chapter we examine this new family of worlds, and finalize the above communication and graduation tools to the assessment of cognitive algorithms as a counterpart of the inferential tools discussed in the previous chapter. Here the distinction between the conceptual model and the implementing hardware goes to vanish so that we will refer for instance to neural networks or quantum bits which precisely implement a given cognitive algorithm. The common features of these algorithms have to be searched in the triadic architecture Computation-Communication-Memory and in general strategies aimed at optimizing the overall objective of this contrivance, having feedback as a precious communication facility. This chapter offers a reasoned summary of many advances in computer science that have taken place in the last four decades. As such, it's a very streamlined summary, which once again is oriented toward delineating compatible worlds in the province of the golden kit. The reader will probably not find new insights; rather, I hope he will enjoy a systematic view of current cognitive challenges in the field.

When we don't know how to solve a problem, we ask around, thus calling upon people to put in share their abilities with us. The key word is *communication* in the specific acceptation of exchange of knowledge between individuals. Thus, while Nature conveys information to us through samples that are in principle equally weighted, people share information that is weighted by the personal knowledge underlying it. This is a category that plays many roles, such as experience, creativity, inspiration, etc. Abandoning the perspective of celestial revelations, ecstasy and other mystical things, we focus therein on the role of experience, which we manage formally through fuzzy sets. Hence, in very sharp terms and with some abuse, besides its technological aspects (relevant and astonishing per se), here communication represents a medium for merging fuzzy sets, whose management is demanded to cognitive systems.

Non uniform weights by experience.

B. Apolloni, *A Few Things I Know About Her*, Intelligent Systems Reference Library 219, https://doi.org/10.1007/978-3-030-94379-0_4

4.1 The Fuzziness Recovery

In the cases where we may describe data through distribution laws, the compatibility of an inferred parameter is exploited in terms of the probability of being correct if we adopt the same inference procedure in a long history of instances of our lives. Nevertheless, we cannot hypothesize our future if we lose the overall vision of our environment, at the basis of probabilities. Thus, we must introduce other ways of exploiting data when we may exploit only *local* information to decide the *plausibility* of a given attribute. The favorite one is the *fuzzy sets* framework, which I introduce as follows.

The role of the synthesis of a long past to render the present comfortable.

In a true essence, in this framework the long term perspective on data concerns the past, in place of the future. Thus, in place of probabilities on what could happen late in the time or the space, *plausibility* synthesizes the knowledge gained in the past by the single individual through his *experience*.

Let us consider a very operational case study. The latest versions of housekeeping appliances are endowed with dialogue interfaces to understand preferences and needs of the users. So, we may expect in the upcoming future a washing machine to be equipped with an interface like in the picture asking us about the next laundry, so as to feature automatically the optimal washing cycle.

src 15

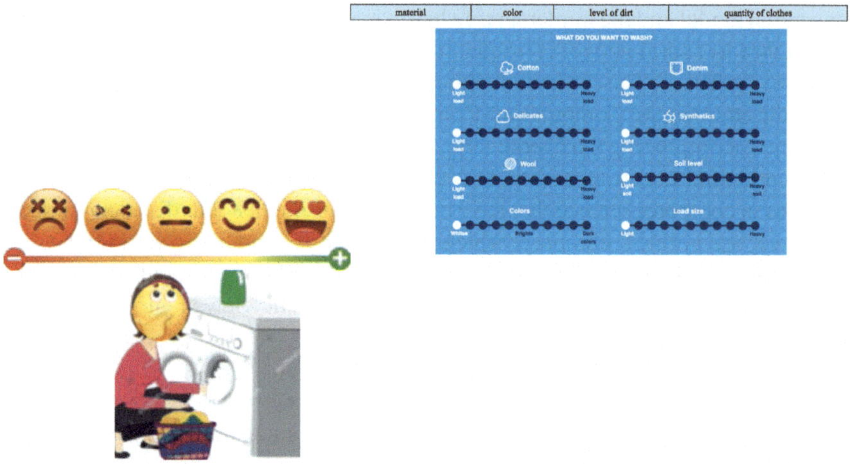

If I can't explain,

In this scenario, if someone asks us why we clicked on the third bullet of the dirt level graduation, we may say that "we felt like it.". And if the same person asks us what we think about the information brought by the third bullet of the dirt level graduation, we may answer with the customary example of the attributes *bald* and *rather bald* as in the figure below.

- While for sure we denote as bald a guy with no hair at all on his head,
- for sure as no bald one with around 200 hairs per square centimeter,
- we will say that a guy with 20 hairs per cm^2 is in an intermediate position, which we may characterize with curves like in the figure.

Those curves describe the plausibility of the selected attributes in terms of membership functions μ_b and μ_{rb} of a guy to the fuzzy sets respectively "bald guy" and "rather bald guy" as specifications of the fuzzy attribute "bald". Analogously for further gradations, such as "not so bald," "full head of hair," etc.

I can rigorously describe,

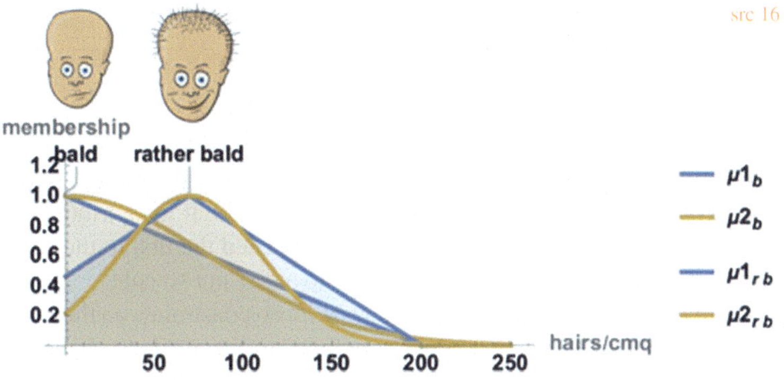

In general terms, a fuzzy set is a set X whose support is in principle the entire real axis (as for unidimensional sets[1]), but each point x of the support is affected by a real number (customary in [0, 1]) denoting the membership degree of the point to the fuzzy set. Like for random variables, we associate to X a set of values and a set of graduations of them, yet with different properties. Different graduations may coexist on the same support, respectively denoting different fuzzy sets that describe the support in a non exclusive way.

The shape of the membership function is exactly up to us; it is the synthesis of the many events of our life that have fostered a given shape. Thus, in principle, it may be of any kind, provided some obvious properties (such as non-negativity) are satisfied. On the other hand, materializing such a shape could require the intervention of psychoanalysts or other managers of the unconsciousness, yet still with no guaranteed results. Thus the effort that is commonly required to the people is just to identify two-three parameters that differentiate families of standard-unimodal curves. In the figure below we show two of them:

through simple *plausibility* curves

- a triangular one $\mu_{1A_{\{a,b,c\}}}(x)$, because of its simplicity, and
- an asymmetric Gaussian-like one $\mu_{2A_{\{\nu,\sigma_l,\sigma_r\}}}(x)$, because of its smoothness; both require only three parameters to be specified.

[1] The entry level option of ours for this instance.

Namely:

$$\mu 1_{A_{\{a,b,c\}}}(x) = \begin{cases} \frac{x-a}{(b-a)} & \text{if } a \leq x \leq b \\ \frac{c-x}{(c-b)} & \text{if } b < x \leq c \end{cases}$$

$$\mu 2_{A_{\{\nu,\sigma_l,\sigma_r\}}}(x) = \begin{cases} \mathrm{e}^{-\frac{(x-\nu)^2}{\sigma_l^2}} & \text{if } x \leq \nu \\ \mathrm{e}^{-\frac{(x-\nu)^2}{\sigma_r^2}} & \text{if } x > \nu \end{cases}$$

In the picture we represented three instances $A_{\{\alpha_i,\beta_i,\gamma_i\}}$ of them, with $i \in \{1, 2, 3\}$, which respectively describe the sets: "small", "medium" and "large", with extra-constrains on the left of $x = 0$ and on the right of $x = 250$. If the attributes concern the number of hairs, in turn these sets could be associated the first to the set "*b*ald", the second to the set "*r*ather bald" and the third to a set "not so bald".

with proper parameters.

Thus, membership functions are local grades with no constraints on their integrals. We may just say if a given number of hairs belongs to the attribute "bald" more or less than another number. If someone ask us: "Why do you give membership 0.8 to $x = 20$," we may answer: "Because I feel like it". Thus the compatibility value of this membership concerns the past and affects the values of the fuzzy attribute.

4.1.1 *Exploiting Fuzzy Sets*

The immediate question arising about memberships is how we can exploit them. There are two aspects, an operational one and a modeling one.

As for the former, *defuzzification* is the counterpart of estimation in this framework. It is aimed to derive special values of the fuzzy set support to be employed in operational contexts.

Reasonably narrowing the fuzzy set support, possibly to a single vale.

- The analogous of the $(1 - \delta)$-confidence interval is the α-cut, which gathers all values in the support having a membership $\mu \geq \alpha$.
- Point evaluations are more variegated, ranging from the
  - "center of gravity" of μ as a counterpart of $E[X]$, to an
  - α-cut limit with α growing up to its maximum, as the counterpart of the modal value of a probability density function, to
  - mean of the extremes of a particular α-cut as a loose counterpart of the median of a random variable.

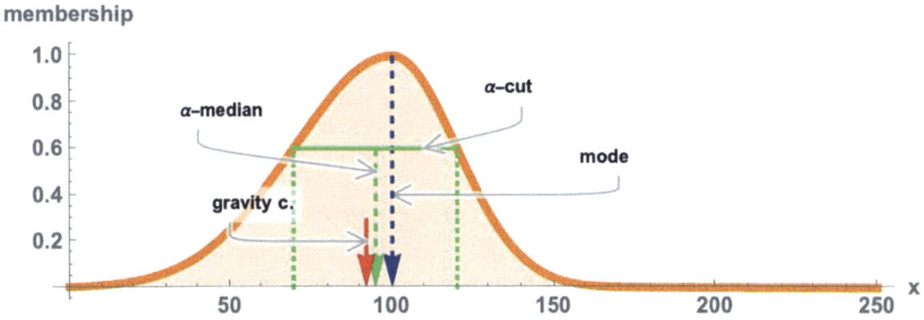

Actually, in the literature we may find many other solutions, for they depend again on the experience of the users, rather than on cogent theoretical properties.

The second question concerns how to compose fuzzy attributes to create complex fuzzy models.

Like with events of a probability space, we look for the basic operations of negation, intersection and union of events, now referring not to the support of the event, which is infinite by default, but to its membership function. Namely, **Combining fuzzy sets.**

negation We obtain the complement $\bar{A}$ of a fuzzy set A by variously negating its membership $\mu_A(x)$. When μ is clipped in [0, 1], the most common[2] negation is

$$\mu_{\bar{A}}(x) = 1 - \mu_A(x)$$

so that the complement of $\mu_{\bar{A}}$ is μ_A, again.

intersection We obtain the intersection C of two fuzzy sets A and B by computing a *T-norm* of μ_A and μ_B. Most common T-norms are

$$\mu_C(x) = \min(\mu_A(x), \mu_B(x)) \quad \text{and} \quad \mu_C(x) = \mu_A(x) \times \mu_B(x)$$

union In a conjugate way, we obtain the union D of two fuzzy sets A and B by computing a *T-conorm* of μ_A and μ_B. Most common T-conorms are

$$\mu_C(x) = \max(\mu_A(x), \mu_B(x)) \quad \text{and} \quad \mu_C(x) = \mu_A(x) + \mu_B(x)$$

[2] Another could be $\mu_{\bar{A}}(x) = \frac{1-\mu_A(x)}{1+\lambda\mu_A(x)}$ with $\lambda > -1$.

We may consider the above operations as a shift from the analogous operations on events, due to the poorer information framework. Namely, in the probabilistic framework, the probability of a composition of events is ruled by the conditional probability formula on Sect. 2.4.4. Thus the connectives $AND(\&)$ and OR (||) with respect to the disjoint events A and B and a third event $E = (E\&A)||(E\&B)$ give rise to the formula

$$P(E) = P(E|A)P(A) + P(E|B)P(B).$$

Hence, (i) in a sample space $\Omega = A \cup B$ and (ii) in the availability of local probabilities $P(E|A)$ and $P(E|B)$, respectively synthesizing observations on E in the two complementary frameworks A and B, the overall probability P(E) is a weighted sum of the local probabilities. Here the weights are the relevances (the probabilities) of the frameworks A and B where local probabilities have been casted, i.e. the probabilities that the local conditions A and B occur. Finally, P(B) = 1 − P(A) (by simple derivation from the probability axioms on Sect. 2.4.1), so that the above operation is a *convex* linear combination (see the figure below)

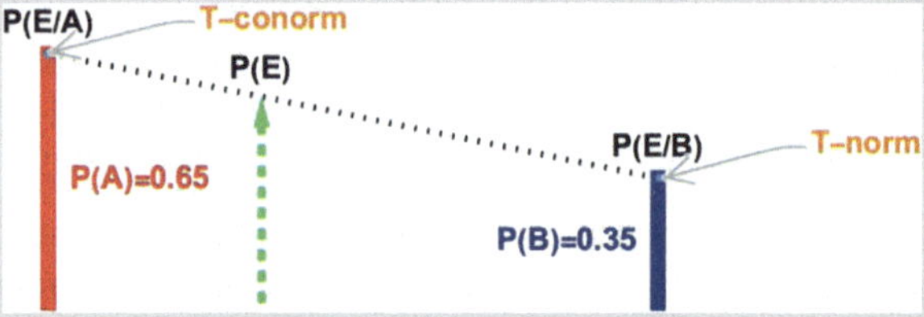

With fuzzy sets we no have sufficient information to compare events of a sample space. Rather we vanish the border of any event, which in principle now spans all the sample space and give a local compatibility of each point inside it with our experience. In this framework, we simply establish that:

- the plausibility of a point to be jointly a member of two fuzzy sets is biased by the lowest of the two memberships of the point to the sets, via a T-norms – as a counterpart of the probability of a conjunction of events.
- Analogously, the plausibility to be a member of any of them or both is drawn by the highest of the two memberships via a T-conorms.

Thus, in the fuzzy framework, in the absence of accurate weights, we come directly to the extremes of the linear combination to denote either the conjunction or the disjunction of the fuzzy sets as surrogates of giving weight 1 to either the less probable or to the highest probable conditioning event. Are these extremes still probabilities? In the lack of $P(A)$ and $P(B)$, this is immaterial for the final exploitation. Rather, it may be a way of identifying the membership functions. Thus, for instance, we may ask a team of m colleagues (any people whose expertise we trust in) about the membership of $x = 20$ to the attribute "bald" and take an average of their answers. The answers may still be degrees synthesizing their experience or they may be binary, hence forcing colleagues to answer "yes/not". In any case, these statistics can be used only as membership degrees of $x = 20$ to the attribute "bald."

4.2 The Communication Recovery

Communication for a virtual globality.

In the lack of an overview allowing us to handle probabilities of the elements of a universe, we may rely on another attitude, namely *communication*, which enables us to potentially reach those elements even if they are hidden to us. Communication is a key feature of our times. Obviously it has been the root of human development since the beginning of time, but we began realizing its ontology, hence abstracting on it, not so long ago.

In the mechanistic vision of the world I remember in my youth relations between people and objects were ruled by the physical contact between items, possibly through intermediate entities such as mails, or possibly with periodical oscillations as in the case of the vibration of music strings. Dealing with interactions not explainable with physical contacts is undoubtedly an intriguing matter calling into question extra-natural phenomena, so very much in the territory of the soul indeed. As a practitioner in physics, I notice that the law of gravity may represent in classical physics a last legacy of the *virtus dormitiva* ancient way of approaching natural sciences[3]: an explanation which merely restates in different words the very thing which is to be explained. Gravity exists, and that's it. We find a clear description of its ruling, which derives from sophisticated speculations of Isaac Newton and give a rationale to a plenty of phenomena in Nature, yet nothing about *why*. Actually, gravity is the oldest form of remote communication experienced by humans. Sound is another one, but it is linked to bodies (molecules) contact.

Communication, like gravitation but more complex.

Like gravity, another *immaterial* medium of communication is represented by electromagnetic waves. It's a favorite medium because of their manageability, so that no person in the world, apart tiny babies and members of very primitive communities, can imagine his life without this medium. My personal view it that communication is precisely what enables us (our brain) to work with virtually overall universes without individually knowing fully their elements, neither as for their number. We could define this framework as an extended fuzzy sets framework where the individual experiences are empowered by communication with others.

[3] From [15]: "Moliere, long ago, depicted an oral doctoral examination in which the learned doctors ask the candidate to state the "cause and reason" why opium puts people to sleep. The candidate triumphantly answers in dog Latin, "Because there is in it a dormitive principle (virtus dormitiva)".

What we exploit of data depends on what we communicate of them.

To consider the variety of information scenarios we may meet in a communication network, let's imagine we are a company having a social network at our disposal, and we want to assess customer satisfaction re three alternative featuring (as for sugar level, flavorings and alcohol content) of a *pop*ular alcoholic soft drink. Hence we ask the network members about their experience consuming the diverse versions of the drink.

src 17

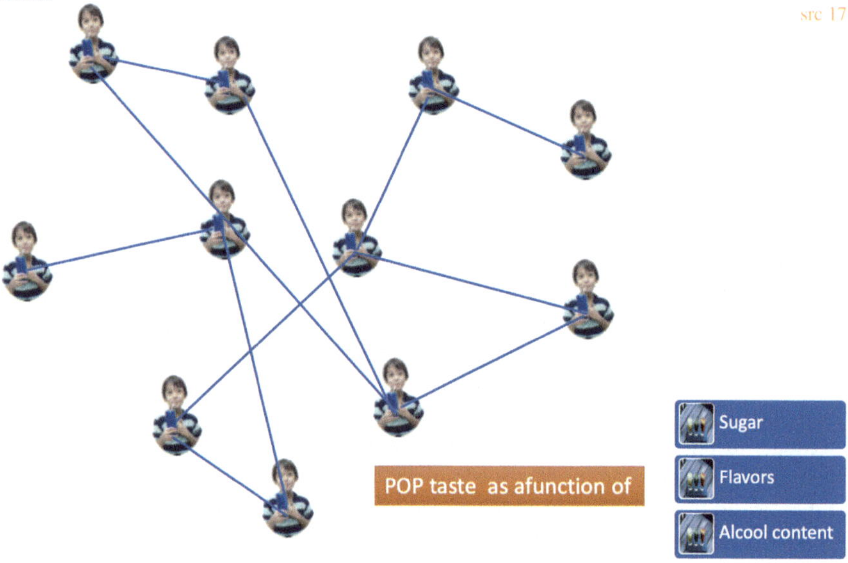

The goal of the company is to partition its production of the drink in accordance with market acceptance. What's the best way to manage the user experience? We list some alternatives and conditions that may affect the experience.

1. binary vote or a more detailed one (for instance, a graduation between 1 and 10)
2. the user is focused on a single variety or considers more than one
3. user reliability
4. number of users
5. representativity of responders versus the target market.

Let's consider the basic scenario where the vote is binary (just for simplicity's sake): each user tests all drink variants and is equally reliable and well representative of the market. In this case the estimates of the selling probabilities of the variants may be simply computed by putting in pool the votes of the users and by computing the frequencies of 1 accumulated on the individual variants. In the next picture we may see how the votes of 20 users merge to compute the probabilities.

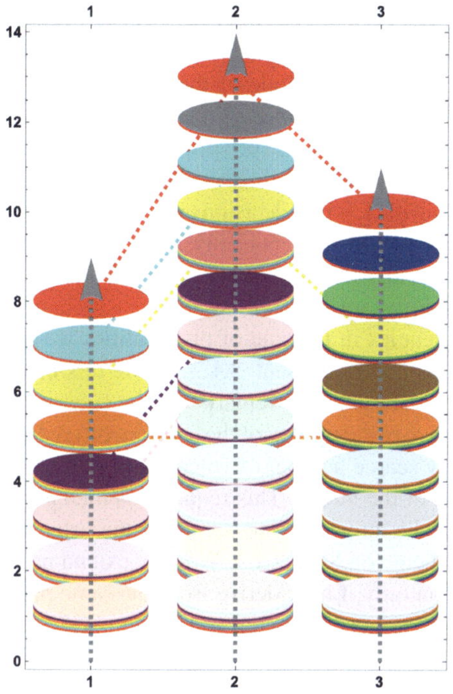

Colors denote the different users; disks appear in the column where the users vote 1. If we move along the vertical dotted arrow we cumulate the 1's expressed by the users on the individual variant (horizontal dotted lines join the positive votes of a single taster). This is local information that we may use to appreciate the favor of the users in regard to the relative variant. Thus we reckon that 8 users out of 20 like the first variant, 13 the second and 10 the third. Hence we have local frequencies $^{8}/_{20}$, $^{13}/_{20}$, $^{10}/_{20}$. At an overall level, the manufacturer knows that out of 60 tasted bottles, only 31 got a favorable vote, and out of these 31, user preference was distributed with the respective frequencies of $^{8}/_{31}$, $^{13}/_{31}$, $^{10}/_{31}$. What changes when we go from local frequencies, in the role of pleasantness grade, to global frequencies, in the role of probabilities, is a normalization leading the global frequencies to 1. But we can normalize only if we know the local frequencies of all variants.

Note that the experiences of the individual user may add up the other ones to generate a graduation, but their frequencies over variants don't add up to generate the overall frequency. However, the mean of the local frequencies may represent a good approximation of the global frequencies when frequencies are taken over similar numbers of voters. This is why the local consensus in a social network may be used to form a global consensus; however, the latter reflects only approximately the global probability of receiving a consensus.

Be cautious to merge local feelings.

Things get worse if some of the above features of the users are lost. Namely:

1. If a user doesn't taste a variant we cannot reverse automatically local frequencies into the global one, since there are some holes in the data to be taken into account. This has the dual consequence of not giving each user a same weight $^{1}/_{20}$ and of

partially losing the representativity of the users since special cases enter into play, where some users were not put in the condition of tasting all variants. It's what would happen if the company includes in the pool of testers people from countries where alcohol is forbidden for *superior* reasons.

2. Different reliabilities of the users, too, imply loss of equal weight of the votes. The last property is generally established in the case of political elections, but may represent a drawback when decisions must be taken by skilled people. The effect on our case study is that, not only locally reckoned frequencies cannot be reversed automatically into global ones, but also the consensus accumulation should be handled with care.
3. To get stable frequencies in the role of probabilities we need large numbers of voters. We spoke about this on Sect. 3.1.2, where this number is also related to the variety of the vote targets. We remark that the consensus generated on social networks, for instance re the quality of the services offered by a hotel, is sometime based in a distinctly scarce number of voters (less than 10).
4. Representativeness of the voters. This requisite is the operational counterpart of the horizontality of the threshold line decreeing the value of a Bernoulli variable according to the scheme on Sect. 2.1.1. We may extend properties of a sample to the entire population only if both derive from the same phenomenon. Otherwise it would be as if the various Witchcraft Acts[4] approved by the British Parliament in the 16th century were to be applied in U.K. today, post Brexit.

The key to distinguishing between the various exploitations of the taster verdicts passes again through the notion of conditional information and probability, like in the previous gray-box. In abstract terms, we assume the outcome space Ω to be partitioned in the events $\{A_1, \ldots, A_k\}$ in the picture below, which are variously covered by the event B. Let A_is exhaustively represent the conditions where B may happen, for instance the producers of the CPUs in the world, and E be the irreversible fault of the CPU before a fixed duration. Hence the event intersection $A_i \cap B$ means the fault of a CPU produced by A_i, and $P(B|A_i)$ is the fault probability *given that* the CPU has been produced by A_i. The *total probabilities* formula

$$P(B) = \sum_{i=1}^{k} P(B|A_i)P(A_i)$$

says that to merge the probabilities related to the single producers we must weight them with the probabilities that a CPU is produced by them.

We may consider conditioning events either the 20 tasters or the three products. In both cases, the equal probabilities of the conditioning events allow us to average the local scores to get final probabilities, provided that those scores may be assumed to be estimates of success (if, in turn, the single votes are representatives of the local population). When this information structure fails, we need the further inference tools we will find in the next section.

[4] The Witchcraft Act is a parliamentary act against Conjuration, Witchcraft and dealing with evil and wicked spirits; it establishes that witchcraft is a crime punishable by death.

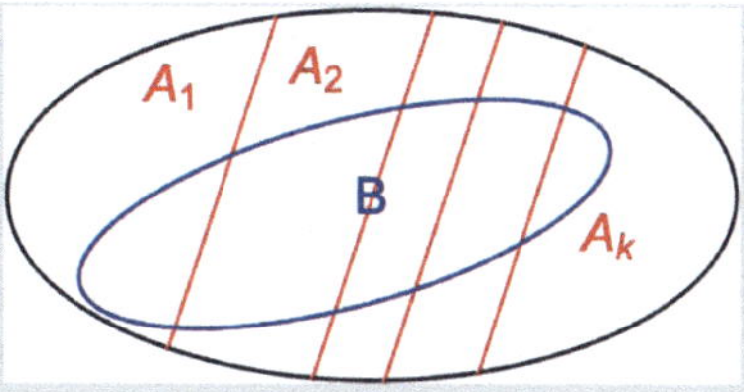

4.3 Cognitive Algorithms

Recalling the essential framework of Chap. 1, we may draw a parallelism with the genotype - phenotype relationship in living organisms, where the former are the inner code of the latter, and the relation between the explanation of a phenomenon and its appearance to our (either natural or artificial) sensing devices. Discovering the former from the latter is assumed to be the core business of human brain, as a reverse engineering task of what Nature constantly does when generating organisms and phenomena in general. Cognitive algorithms are computational tools intended to replicate or even improve this activity with the prosthetic support of a computer. In that, they may represent the latest facility to fruitfully exploit available data to understand phenomena.

Knowing is understanding, quite a *natural* functionality of our brain.

Per usual, we have two ways of dealing with a long sequence of events of a history (for instance of a skilled shooter or an improving football team) to avoid recording it as is in our memory:

- either we may simply group the event according to regularity features, or
- we may deepen the structure of those events, for short, the algorithm which generated them (like the first conceptual task teachers ask us to perform in elementary school: summarize a story using our own words).

The Sampling Mechanism $\mathcal{M}_X = (Z, g_\theta)$ follows the second path, where the data genotype is exactly the explaining function g_θ (to be put in input to a general purpose computer as an artificial Nature machine). In particular, the Universal Sampling Mechanism $\mathcal{M}_X^* = (U, g_\theta)$ offers the solution $g_\theta = F_X^{-1}$. However, if we don't know F_X we don't know its inverse as well, but even if we know F_X it may be that we have no closed form for its inverse, hence we need numerical methods to manage it.

An evolute sampling mechanism:

plus ancillary observations as the input to the explaining function.

Moreover, the events we normally face are not independent; on the contrary, there are more or less complex structures gathering them, that we want to perceive. Cognitive Algorithms both override drawbacks on F_X^{-1} and host data structures by moving the problem upstream. The framework is the one of computational learning, so that, like in the regression problem, g_θ generates samples on the basis of both hidden seeds $\{\zeta_1, \ldots, \zeta_m\}$ and independent variables $\{z_1, \ldots, z_m\}$, where the latter here represent the *primers* of the sample.[5]

Unknowledge as the main seed. Guessing g_θ within a family of general purpose functions, by learning the free parameters θ of the family.

The starting point is again the sentence *I know I don't know*. Here, the uniformity of the unawareness concerns the function explaining the sample. Thus, we may imagine having an empty blackboard (*tabula rasa*), on which we want to write the expression of the explaining function just by guessing if from the observation of a sample of Z and the related sample of X. Obviously, in these terms it is a matter of diviners. We may make the task easier by anchoring it to a general skeleton of a function whose details have to be inferred from the observed data, something like what happens with the *canovaccio* in the Italian Comedy of Art.[6] Our preferred skeleton is the artificial neural network, where we may decide in the advance the architecture, for instance the multilayer perceptron on Sect. 1.3, but then we must guess its parameters from data, as sketched in the "discovery box" below. To cope with the usual representation of the task, in the picture we move z_i's from the set of parameters θ to the input of the explaining function, while we avoid mentioning the seeds ζ_i.

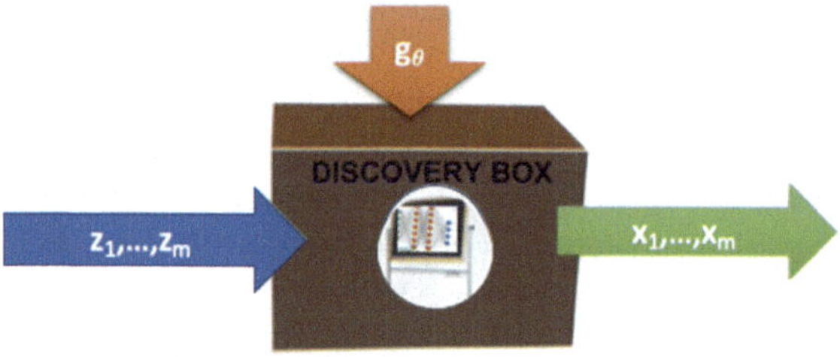

A lot of inventors,

Cognitive algorithms represent a vast prairie where any type of cattle may search for food and receive rewards. In principle, there are no hard constraints and we may invent any cognitive rule, possibly emulating human behavior or even the one of insects like ants or bees. In practice, even though some people adopt the cognitive rules of the NoVax movement, only a few rules survive and spread through the scientific community. And there is a rationale to that. In particular, *surviving*-cognitive-algorithms share four features:

a few surviving algorithms, which share: incremental paradigm,

1. the simplicity of the learning algorithms. In principle, the dry learning activities should be within everyone's reach, as in the case of babies. Thus incremental learning[7] is the most poplar paradigm, where, so wanting, we may introduce sophistication in the determination of the stepwise increment on the parameters under question, but the gradient descent remains the first option.

[5] Hence we deal again with an explaining function, where the explanation goes from known inputs to known outputs, having random seeds in the background.

[6] A list of acts and scenes where the details are left to the improvisation of the actors.

[7] See the footnote on incremental algorithms on Sect. 3.3.2.1.

2. sub-symbolic management. Like for neural networks, learning per se is a pre-scientific activity which does not allow for formal results. A function has been learned if the learning machine (the *canovaccio*) produces correct outputs, even though we cannot describe the function via a symbolic expression. **sub-symbolic target,**
3. communication between atomic learning processes. With reference to an artificial neural network as the mother of all learning engines, the atomic processes are carried out by the single neurons, whereas wise communication protocols wrap up the efforts of the singe neurons to get an overall learning machine. **local out-puts' broad-cast-ing,**
4. the poorness of the theoretical background, so that the quality of learning is checked in a quite empirical way by training the machine on a set of the examples (the training set TRS) and checking the correctness of the learned function on a further set of examples (the test set TTS), thus assessing its *generalization* capability. Possibly, a third set of examples is used at runtime to control the over-fitting phenomenon, which occurs when the learned function is overly adherent to the TRS so that it misses some general features of the function that would allow a satisfactory adherence of it to the TTS as well (see footnote 11 in Sect. 5.2). Thus the TTS is used to calibrate the training duration and a further set of examples, the validation set VDT, is used to check the generalization capability of the learned function. **empiri-cal checks.**

These four characteristics render the cognitive algorithms so essential that some standard implementations enter into the literature and into the market as well, to be exploited drop-down-menu style. **Simple, not stupid.**

Be careful, this does not mean that learning a function is nowadays a task for kids. There are many issues where expertise and knowledge count. For instance:

- The identification of the target of the incremental learning, i.e. the optimization goal, which we will discuss in a next section. **Proper target,**
- The preprocessing of the data. The art of teaching consists of presenting the tasks to pupils in very simple terms. This is the province of the preprocessing the input of the algorithm. For instance, a popular preprocessing of images is the convolution of the pixels. **pre-processing,**
  In general, we may split the learning task in two, a first part where we inject all available knowledge of ours to simplify the core of the learning task, followed by the dry learning of the residual function uniquely from the *simplified* data. At this stage we may see the input of the function like a set of random variables, since any reachable distinguishing feature of them has already been drawn to produce the output of the function. Those inputs act as the primers of the output via a sub-symbolic explaining function.
- The conditioning of the training process with proper ancillary inputs, for instance in the thread of the *generative approach* to statistics, as in scrolling the process away from local minimums. **direct-ing the train-ing.**
- The instantiation of new network layouts and training methods, when necessary.

4.4 Unawareness Management

We may survey the trend leading to the spectrum of the cognitive algorithms from two perspectives: the **completeness of the available information** and the **goal of the algorithms** (see the picture below).

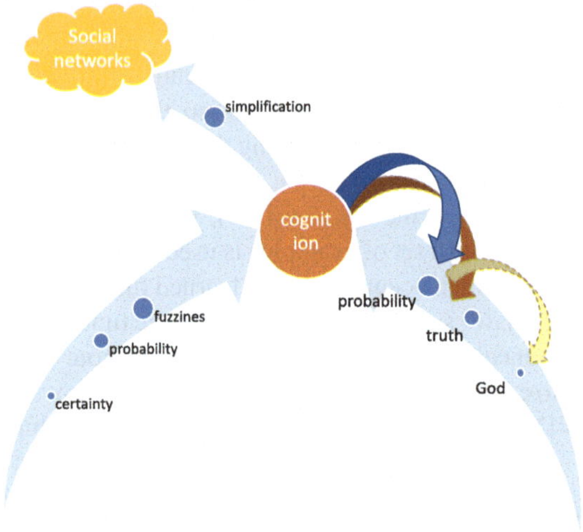

4.4.1 The Information Path

As for the former, we have already analyzed the *probability* and *fuzzyness* steps, both aimed at establishing a common language in spite of noise and uncertainty that necessarily moved us away from the first step, *certainty*. Cognitive systems represent the current evolution of regression models. The common target is the function explaining the transition from the seed to the variable in question.

g **elementary → we understand what we are explaining;**

- The first family of explaining functions was the family of linear functions, as the mathematical counterpart of drawing a line roughly joining a set of points in a plane. For instance, the graph below shows the times to reach by feet a set of destinations as a function of their distance from our current position (typical values supplied by GoogleMaps).

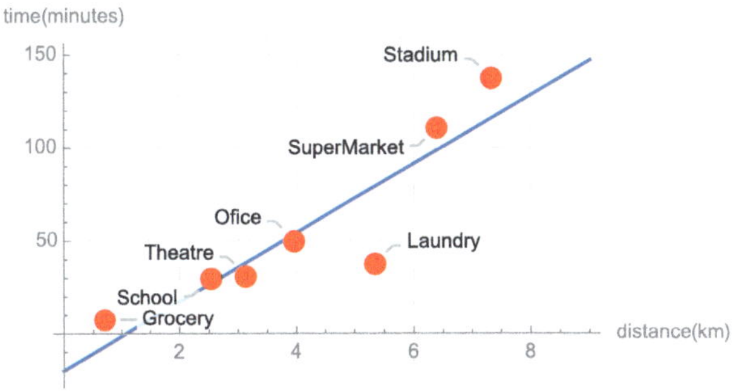

Drawing a straight line fitting the points is a very elementary operation thanks to the elementariness of the fitting function. Drawing it with some optimality features, such as minimizing the vertical shift from points to the straight line, is demanding for some algebra yet is normally achieved.

- If no one/no evidence told us that linear fitting is the best we can do, we are allowed to check more complex functions, for instance a quadratic one, as in the picture below.

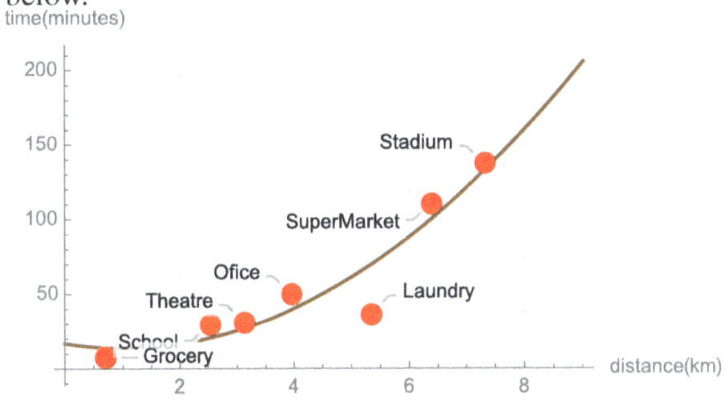

We might involve other variables, such as atmospheric temperature – as a marker of affecting phenomena such as traffic jam, to explain the observed run times, so obtaining pictures like the one below.

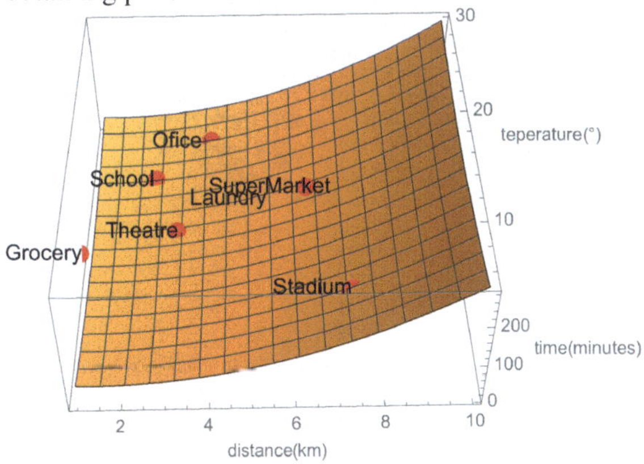

g complex → we explain even accurately, but we don't understand.

- Thus, why not any kind of function? This is the new deal of the cognitive systems. We rely on a base function that is in principle susceptible to regressing any explaining function, the forementioned *canovaccio*, provided we are able to identify its free parameters and provided they are in a sufficient number. Our preferred canovaccio is a neural network, preferably organized as a multilayer perceptron. The mentioned Cybenko theorem assures us about the full potential of this base function, as long as we have a sufficient number of free parameters. The problem is how to identify their values. This is the goal of the learning algorithms, which have recently been revived through their implementation on the Deep Neural Network, a perceptron with many layers, each one possibly specialized on partial subgoals of learning. In essence, the difference between regression procedures and cognitive systems lies in the relevance of the parameters to be inferred. In the regression models, we have a symbolical description of the regression function to be completed with the inferred parameters. In cognitive systems, we have no idea of the symbolic shape of the function underlying the data, if any; rather, we have a general purpose function endowed with many parameters, like in the case of neural networks, so that they are the values of the parameters that implicitly shape the underlying function.

Provided it works, any biological system may prove inspiring.

- Obviously, neural networks are not the unique solution. Another consistent thread is represented by the bio-inspired algorithms, mainly in the branches of genetic algorithms and swarm intelligence. Genetic algorithms aim to modify the architecture of the cognitive systems, as occurs with the evolution of the genome to increase the survival capacity of living organisms (recall what we discussed in Chap. 1). Swarm intelligence mimics the optimization strategies implemented by swarms of insects and small animals in meeting survival challenges (think of ant colonies in search and recover of food).

src 18

4.4.2 The Goal Path

It depends on what we are looking for:

The second path mainly reflects my opinion and may prove naive as I'm not a sociologist.

- Sharply, we may figure that the initial goal of the human brain was drawn by the explanation of phenomena in terms of the will of God, which, at the end of the day, represented both an abstraction and a systematization of them, as a primary remedy to unawareness.

  - a divinity ruling our life,

- Then, people began replacing God's will with their own minds. It was not an easy transition, mind to the Galileo abjuration,[8] that inherited, however, the idea of an absolute truth to be discovered and defended. Classical physics represent a *Summa* of the mind's findings up until the beginning of the last century.

  - the mind ruling our life,

- Probability, within this thread, held a way of bypassing many holes left by physics re the certainty of laws and their effects (again, a biased vision of mine). So, laws are universal, but effects may be uncertain. But uncertainty too may be ruled, ergo we have the birth of probabilistic models and statistical tools for implementing them on observed phenomena. Actually, the entire classical statistic may be read according to the following scheme:
  In principle there is a distribution law with which Nature generates values of some parameters. Actually, the distribution law identifies with the entire (infinite) population of values, as in the neverending strip on the clouds in the next picture, for a binary variable (Bernoulli distribution). We, aka poor humans, observe from the earth and may look only at a finite sample of the population. Nevertheless, we want to infer something about the population (the parameter p of the Bernoulli

  - eternal rules to discover,

[8] "I, Galileo Galilei, son of the late Vincenzo Galilei, of Florence, aged seventy years, being brought personally to judgment, and kneeling before your Most Eminent and Most Reverend Lords Cardinals, General Inquisitors of the universal Christian republic against heretical depravity, having before my eyes the Holy Gospels, which I touch with my own hands, swear that I have always believed, and now believe, and with the help of God will in future believe, every article which the Holy Catholic and Apostolic Church of Rome holds, teaches, and preaches. But because I have been enjoined by this Holy Office altogether to abandon the false opinion which maintains that the sun is the centre and immovable, and forbidden to hold, defend, or teach the said false doctrine in any manner, and after it hath been signified to me that the said doctrine is repugnant with the Holy Scripture, I have written and printed a book, in which I treat of the same doctrine now condemned, and adduce reasons with great force in support of the same, without giving any solution, and therefore have been judged grievously suspected of heresy; that is to say, that I held and believed that the sun is the centre of the universe and is immovable, and that the earth is not the centre and is movable; willing, therefore, to remove from the minds of your Eminences, and of every Catholic Christian, this vehement suspicion rightfully entertained toward me, with a sincere heart and unfeigned faith, I abjure, curse, and detest the said errors and heresies, and generally every other error and sect contrary to Holy Church; and I swear that I will never more in future say or assert anything verbally, or in writing, which may give rise to a similar suspicion of me; but if I shall know any heretic, or anyone suspected of heresy, that I will denounce him to this Holy Office, or to the Inquisitor or Ordinary of the place where I may be; I swear, moreover, and promise, that I will fulfil and observe fully, all the penances which have been or shall be laid on me by this Holy Office. But if it shall happen that I violate any of my said promises, oaths, and protestations (which God avert!), I subject myself to all the pains and punishments which have been decreed and promulgated by the sacred canons, and other general and particular constitutions, against delinquents of this description. So may God help me, and his Holy Gospels which I touch with my own hands. I, the above-named Galileo Galilei, have abjured, sworn, promised, and bound myself as above, and in witness thereof with my own hand have subscribed this present writing of my abjuration, which I have recited word for word. At Rome, in the Convent of Minerva, June 22, 1633. I, Galileo Galilei, have abjured as above with my own hand." https://todayinsci.com/G/Galilei_Galileo/Galilei_Galileo-QuotationsAbjuration.htm.

variable, for example), where our inference scheme is based on hypotheses on the population that have to be checked and accepted with some risk, or on properties that are extremal (worst case) with respect to the hypotheses range.

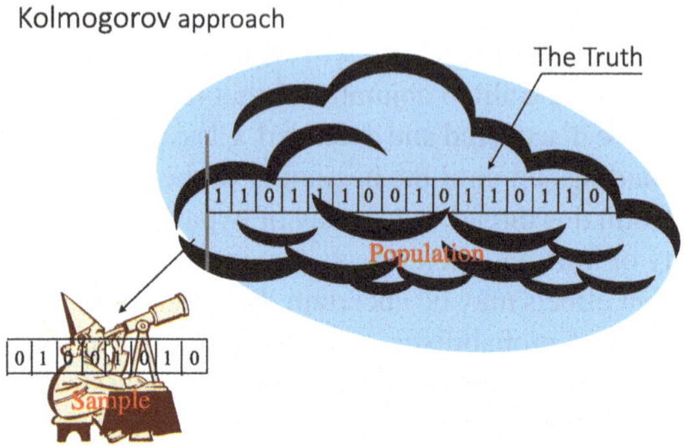

In this framework, the shift of the sample mean $\bar{x} = \sum {}^{x_i}/_m$ from the *true* p is order $^1/\sqrt{m}$. So, if my son tells me that, although his team won 6 matches out of 20 he hypothesizes the probability of a win for the next one to be 0.5 (hence the same as for a loss), I cannot reject this hypothesis with the standard risk of 5% used in forensic, neither if the wins were only 5 with the (lower) standard risk of 1%.

by testing hypotheses on them.

Indeed, the decision engine allowed by this approach is as follows: we hypothesize that the distribution law in the sky has some features, for instance a Bernoulli distribution with $p = 0.5$, and check whether the observed sample contradicts this hypothesis. For instance, a frequency of 1 equal to $\phi = {}^1/_{20}$ looks *far* from this hypothesis, since the probability of the related random variable Φ deriving from that Bernoulli distribution is such that $P(\Phi \leq {}^1/_{20}) \simeq 0.00002$, which grows to 0.02 and 0.06 for frequencies ${}^5/_{20}$ and ${}^6/_{20}$ respectively. The decision about which probability t decrees the contradiction, i.e. I reject the sentence of my son – let' us call it hypothesis H_0– if $P(\Phi \leq \bar{x}) < t$ is anchored to the negative consequences of being too loose. We range between the extremal condition of setting $t = 0$, which entails that we always accept (not reject) the hypothesis $p = 0.5$ with no tool to realize when this is not true, and the one of setting $t = 1$, which entails always rejecting the hypothesis, for whatever sample ϕ.[9]

With this engine, imagine we have bought a home in a country and after 10 corings you discover that 2 carrots denote soil pollution. If you sue the seller, the justice may establish that there is no problem since we could not reject the hypothesis of a probability of pollution in the inspected area less than 0.09 with a *risk* 5% and

[9] The engine outcome is synthesized by the following decision table [31]

dec / truth	H_0	$\bar{H}_0$
accept H_0	$1-\alpha$	β
reject H_0	α	$1-\beta$

where α and β are, respectively, the probabilities of rejecting our hypothesis H_0 when it is true and accepting it when it is false.

even 0.05 with a risk 1%.[10] That's for court decisions. For sure, we would not tell our son to buy a home in the same country. Indeed, let's assume the pollution to be spread uniformly throughout the area so that here and there some carrots might denote pollution (imagine a detergent spill in one spot, a toxic trash deposit in another, and so on… it can happen). Then, following the same reasoning as with my son's football team, finding 2 polluted carrots out of 10 entails a distribution law of the variable P, which denotes the probability of finding the next carrot polluted (the one corresponding to the place of the candidate home of my son), as in the picture below, ranging from 0.03 to 0.5 with a probability 0.9 and a mean value about 0.22.

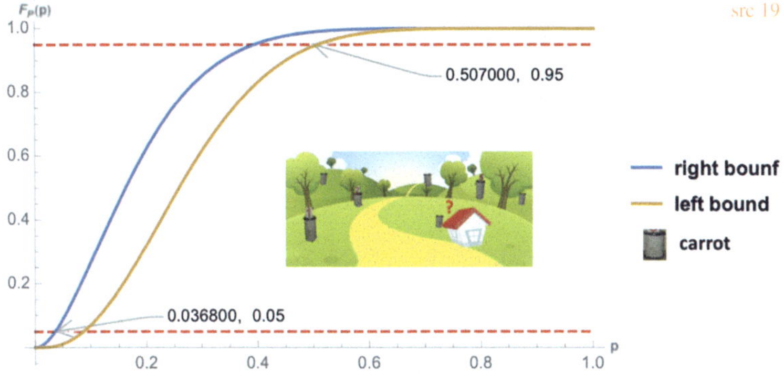

Hypotheses on the shape of the polluted area, for instance belonging to the class of circles as a result of pollutant spillage, like on Sect. 3.3.1.1, may help to refine the inference in terms of learning the polluted area.

- Quantum mechanics may be viewed as a physical modeling companion of the cognitive systems. Again, communication between individuals (between the particles or even inside them) is the key feature. It breaks up their non overlapping features so that the probabilities of their superposition states no longer add up but their square roots, the amplitude, do. Consider for instance Young's famous double slit experiment where a monochromatic light passes through two narrow slits and hits a distant screen. We observe on the screen bright and dark fringes as shown on thew left side of the figure below. They realize an interference phenomenon due to the superposition of the light waves originating from the slits. Due to their different phases, in turn due to the different paths from the slits, superposition might enhance the light in some points and darken it in others. Looking at the right side of the figure, we realize that the light intensity is the result of photons emitted by the source bombarding the screen, so that the intensity distribution coincides with the probability density of the photons along the length of the screen. Should one of the two slits i be obstructed then this distribution is the corresponding P_i

- rules still eternal but non univo-cal.

[10] Note the ambiguity of the notion of *risk* that is totally in favor of the seller. The questioned risk, in fact, is the probability α in the previous footnote: it denotes the probability that the rule:"reject H_0 if the number of polluted carrots is greater than 2" may cause a false rejection when the hypothesis is true. Correctly, this risk is less when the hypothesized pollution is less but concerns the seller's risk.

that we denote as the square of a corresponding ϕ_i which we denote as amplitude. With both slits open the distribution is P_{12} which, because of the mentioned wave superposition, is related to the single P_i through $P_{12} = |\phi_1 + \phi_2|^2$. Hence, with this superposition it's the amplitudes, not the probabilities, that add up.

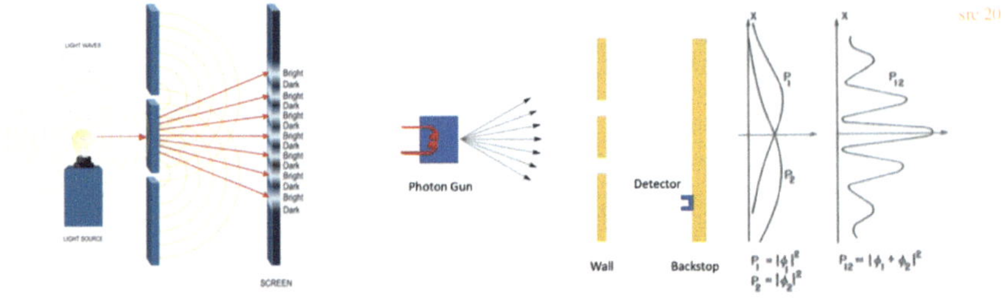

Thus, amplitudes play a role similar to that of membership degrees in fuzzy set theory. They are features that are local to the single particle[11]; but they are richer as well, as they rule the entanglement between a pair of particles. Finally, the cognition approach: "no matter why, whatever works" stems from the relatively large set of postulates of quantum mechanics.

The particles' gossip.

Roughly, we may imagine the particles to be like people who decide to join a party. Each one opts for locations where he can meet people with similar moods and opinions, so as to have a lovely evening. In the absence of communication people would meet in some random way, and those with opposite mood would fight and stay in a black temper. Communication prevents this drawback. Thanks to chats and other types of communication, people may get a clue in advance about the attitudes, opinions and ideas of prospective acquaintances. This allows them to opt for the proper encounters, so that the resulting happening leaves no one in the black zone and moves all people in the brilliant temper zone.

Actually, the physics of the two slits experiment is much more fine and complex, so that Ananthaswamy concluded on 2018 that "physics has yet to complete its passage through the double-slit experiment. The case remains unsolved.".[12] I simply point out that mobile phones are not the exclusive tools of people; special versions of them are available also to elementary particles.

- Undoubtedly, the juxtaposition of quantum mechanics and cognitive systems may appear somewhat daring; thus I hear a disappointed roar of lions. However, let me try to tame them with a few considerations.

---

[11] It is intriguing to note that the same exponent 0.5 is successfully used to superpose states in the well-known algorithm Fuzzy C-mean.

[12] In the book "Through two doors at once: the elegant experiment that captures the enigma of our quantum reality," Dutton, New York.

- Both environments share multiverses, communication and diversion from classical probability descriptional tools.
- The common starting point is the "sensible experiences" claimed by Galileo. Then, in quantum physics those experiences lead to a unique physical meta-world that includes infinite varieties of multiverses in a superposition state. In our approach, for whatever physical world, we reason on which, among compatible worlds, may prove well suited from an operational perspective (say, our decisions). Thus, even though quantum multiverses locally collapse after our observations, the worlds compatible with those observations do not collapse. This could open an apparently unrecoverable drift between the two environments, if the collapsed measures and the surroundings would eliminate any uncertainty. This is not true in general, so that compatibility issues emerge once again.
- All without taking any ecumenical tone, I notice that synergies between the two environments appear frequently in the literature. Apart from the numerous works by physicists in the field of neural networks and fuzzy sets, I may quote an old paper of mine on "Quantum Stochastic Optimization" which is currently referenced [8]. In the occasion of writing it, I had the good fortune to of cooperate with a very skilled quantum physicist. From this merge of physics and information theory competences, a method arose to solve combinatorial optimization problems which is substantially sub-symbolic but rooted on tunneling effect.

Quantum mechanics and cognitive systems are not so far from each other.

4.4.3 *The Place of Cognitive Algorithm*

Cognitive everywhere.

Cognitive algorithms are everywhere, at least according to the statements of the sellers. They go hand in hand with Machine Learning facilities, for instance in our car, but even in our blender or other personal device. They represent a latest innovation in the scientific thought but not its ultimate wrapping out. They constitutes a way of exploiting information that is complementary but not alternative to the other ways mentioned in the information path of the synopsis picture on Sect. 4.4. For instance, along my personal path I implemented them:

- just from scratch in the diagnosis of epilepsies on the basis of a long patient record. It was the largest neural network I ever managed, my earliest implementation as well [5],
- in cooperation with symbolic knowledge, for instance to control the attitude of a geostationary satellite [11],
- in conjunction with some *intentional* mechanisms in some case studies on networks of mobile neurons [7],
- on the top fuzzy rule systems responsive to user feedback, for instance to finetune a bread-making machine [6].

Algorithmic Inference, when appropriate.

However, I relied directly on Algorithmic Inference when facing even complex tasks, such as checking the quality of chip wafers [13] or forecasting epidemic evolutions [10].

As a paradigmatic suggestion, don't train a cognitive system to perform the sum of two integers. Maybe you will succeed after some effort and with some approximation, but any rudimentary calculator can do it better.

src 21

Pure sub-symbolic	**Hybrid**	**Mobile neurons**	Neurofuzzy
	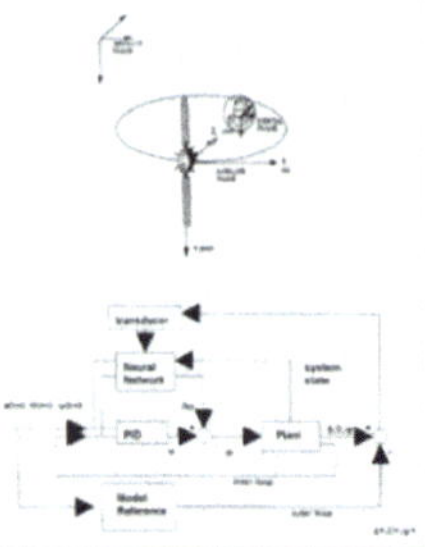	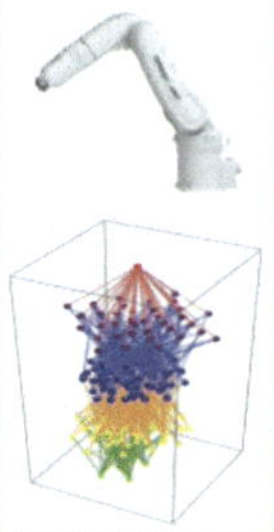	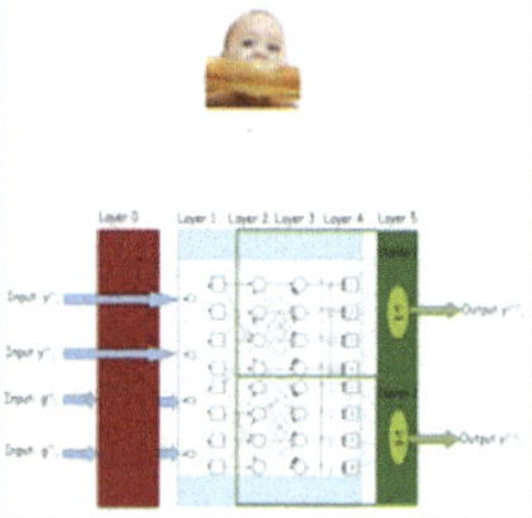

4.4.4 *Continuation Paths After Cognition*

In a very sharp way, let me connote two continuation paths, the one toward a simplification commodity, the other toward a further understanding of nature. Both paths derive from ACO modes of implementing the data–model loop that we will discuss in the next section.

Not always simplification relieves incompetence.

- **The simplification commodity**. *I know I know nothing* is a curse that results as compelling as the situation we're facing is complex, rather, as we have intellectual tools to realize that we are facing complex situations. This may either pose a challenge to us or be a cause of frustration for us. In the second case, people tend to override the frustration by denying the complexity they meet and claiming a need for simplification. In this respect, the claim in one of Aesop's fables: Ἀλώπηξ καί σταφυλή (Sour Grapes! ")[13] appears very up to date. Rather than tackling the difficulty we say that the matter is unduly difficult, and we demand some easy solution. And that's precisely what social networks usually give us. Since we know we don't know, we ask the network. And since we like to know what we expect to know, we ask social networks that are in sync with our opinions. Furthermore, since answers to the questions are given on the basis of their consensus within a community, we fall in the likes' *democrature*. For instance, the picture below shows a partial graph of Facebook pages about vaccination produced on 2020.

[13] Aesop's Fables.

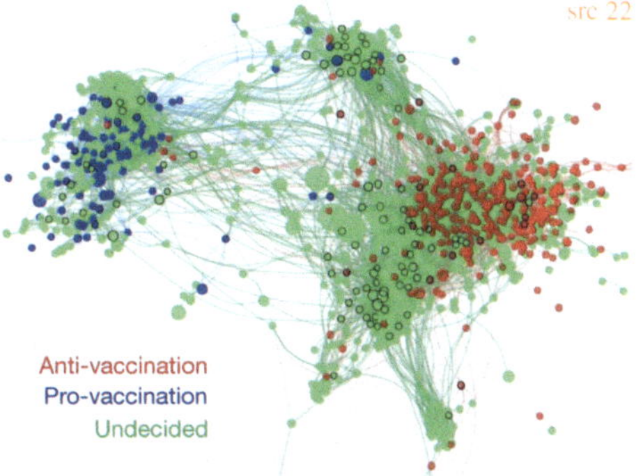

The great diffusion of red nodes representing anti-vaccination clusters evokes a disquieting cultural phenomenon that I call the *Simplification Conspiracy*. It represents a bad answer to the question: *Can a group of improvisers replace a true expert?*[14]

Obviously this is the bad side of social networks; a good one is possible, though we don't always meet it.

- **The further understanding challenge** In the opposite direction, scientists are prompted by the need to understand what they receive from outside their brain. Actually we may put under the wide hat of Artificial Intelligence (AI) what is provided by computers at any level of information and of goal paths. In a figurative way we may say that the current goal is the whitening of any black boxes embedded into the AI tools.

Rather, it helps to approximate sub-symbolic functions with symbolic expressions with affordable complexity.

[14] Note that *ensemble learning* is a successful strategy that has emerged in the last two decades. It is based on a wise synthesis of the results of weak learners [18]. For instance, we may decide to train a set of randomly initialized two layer perceptrons either on the entire training set or on special subsets of them. They should work like experts in different fields of medicine who are called to a patient's bedside to agree on his diagnosis. General agreement rules for weak learners are the average output of the learners, in the case of continuous outputs, or the majority voting in the case of discrete ones (as with the likes reckoning). The divide between improvisers and weak learners lies in how both categories have been trained, if any.

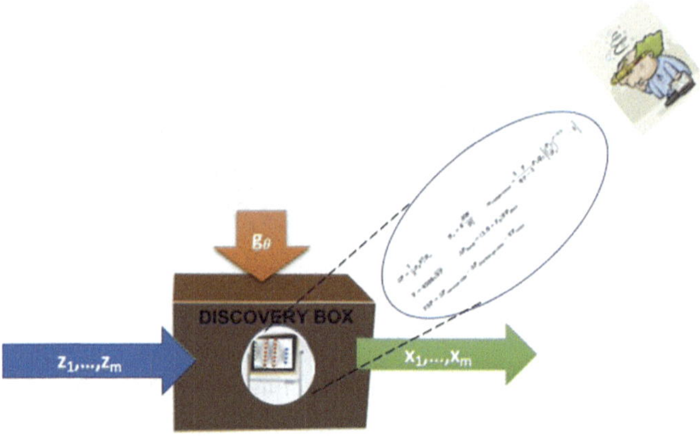

The compatibility notion is of great help:

- we start from a complex function underlying the solution of a given problem, then we totally abandon an explicit model for coping with this function. Hence we surrender to the condition that the desired solution is at least in part sub-symbolic. However, scientists don't give in to the idea that we can recover this function with a wise approximation that is both understandable – since made up of symbolic functions that are easy, endowed with a short number of clear parameters and fed by a small number of variables – and compatible with their concrete experience about the phenomenon the function refers to. This is the vein stemmed by the brown back arrow in the picture on Sect. 4.4.
- we don't give up to the idea of tossing the robustness of the interpretation in terms of statistical properties that can be checked on them (This is the vein stemmed by the blue back arrow on Sect. 4.4). Thus, since nineties techniques have arisen to *translate* trained neural networks into elementary formulas to be understood and agreed upon by those in charge of their usage, a goal commonly denoted as rule extraction from the network [4] and evoked as "passing from synapses to rules" [9]. In particular, major attempts were made to *decompose* the network into its nodes/neurons within a possible simple architecture, so as to read them as atoms of symbolic rules, (for instance OR, AND and NOT gates). The picture below, drawn from [39] describes the translation of a trained perceptron into a series of logical rules. Current approaches will be discussed in the next chapter.

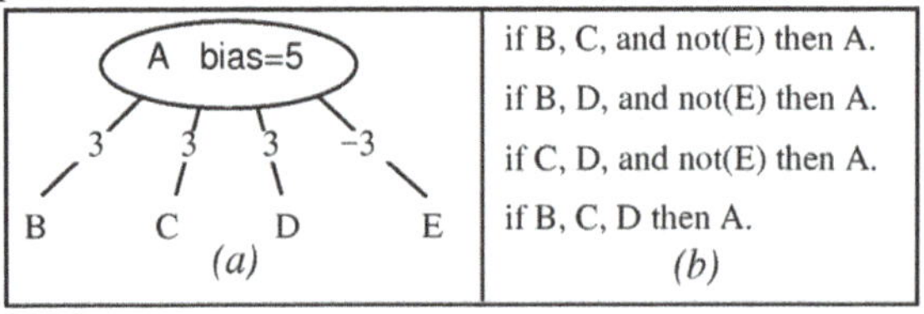

In the figure on Sect. 4.4 there's also a yellow back arrow. But it refers to the conventional approach to probability sketched on the left part in the figure on Sect. 3.6.1, so we contoured the arrow with a dashed line.

4.4.5 And What About Quantum Computing?

This is a tricky question that leaves me puzzled. I cannot avoid mentioning quantum computing among the continuation paths. But I haven't carried out serious research on the field, thus I cannot list it among the few things I know about Her. However, some considerations from the information management perspective come strongly to me, which I want to share with the reader.

The plus of a quantum bit (Qbit) B with respect to the random bit B is the introduction of the $\pm$ sign affecting the probability amplitude of the former. Thus, both B and B may be thought to derive from the experiment on Sect. 2.1.1 and related formula (RandomBit)

The addition of the $\pm$ sign makes the difference.

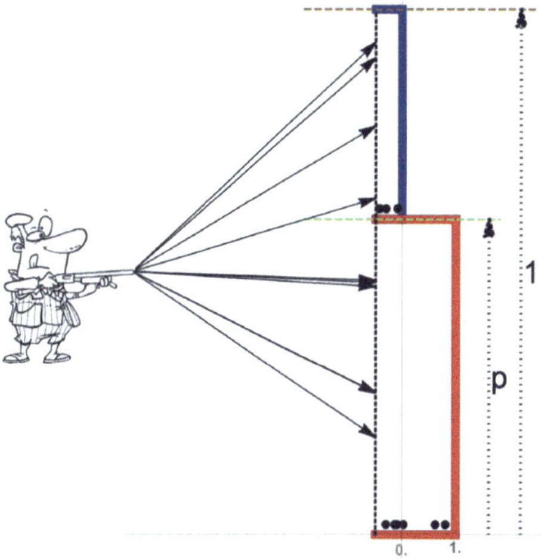

In the picture above, the dotted green threshold line divides the shooter shots between those falling under p, thus getting value 1 exactly with probability p once the shot result has been measured through the caliber of width p and falls in the red slot, and the remaining ones, which get measure 0 after the caliber and fall in the blue slot, with probability $1 - p$. Qbit does the same, getting measure 1 with probability b^2 and measure 0 with probability $a^2 = 1 - b^2$, where a and b represent the amplitude of the measures.

The difference lies in the management of the relations between bits. The Qbit may communicate in an efficient way with other Qbits through the sign of a and b that may be both positive and negative. This entails that the joint values of two Qbits may disappear when they enjoy amplitudes with same absolute value but a different sign, thus entering an interference phenomenon in the same thread as the one considered on Sect. 4.4.2. We may say that the two values of the bits collide, like in the case that another shooter deviates with his bullets the ones of the former when jointly addressed to some pairs of the colored slots.[15] More in general, the measure of one

It supports the interaction between the Qbits,

[15] An event with non-zero probability if some links exist between the behaviors of shooters in spite of their individual randomness.

Qbit may determine the 1s of others, an effect that is denoted *entanglement* and may correspond to an elementary operation of a traditional computer.

which, if implemented jointly on all configurations, figures an extremely powerful computation.

Things are more complex and refined as well. However, within these broad considerations, we may say that collision is a rudimentary form of communication, which is quite different from the elastic bumps of gas particles (the latter like dodgem cars in the moon park). The difference lies both in the bumping effects on the particles, because of the amplitude sign, and in the communication range (and related effects) that may involve all bits jointly. The broadcasting of a signal is not surprising to us in the telecommunication era; rather, the exciting point is that the receiving stations are all the combination of states of the involved Qbits which are in number of 2^n, if the Qbits are in number of n, hence an enormous number even for $n = 50$. This is the quantum parallelism that would solve relevant computational problems, which essentially should change the borders of the manageable computational framework, provided we would be able to implement in hardware the broadcasting of finely controlled communications.

Quantum computers represent a scientific challenge to be tenaciously pursued, albeit with the consciousness of the feasibility limits we are tossing and of the eventuality of attaining only local solutions that, like with parallel computing, may prove suitable in narrow application fields.

4.5 From Data to Model and Back: The Eternal Golden Loop

Observe, elaborate, and then observe better.

Let's consider the loop shown in the figure below. With no pretense of exhaustiveness, it synthetically puts in a functional relation two pillars of he scientific thought: data and models.

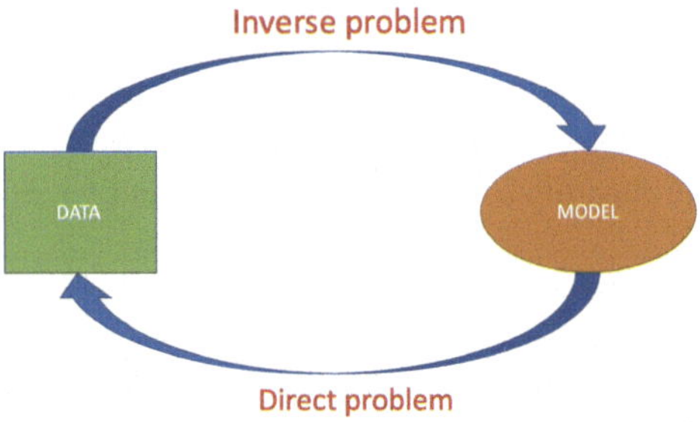

During my engineering studies at University fifty years ago, my education was oriented toward solving the inverse problem in very elementary instances, as a way to get a synthetic description of the phenomena on hand. This was, for instance, by realizing basic properties of elements (e.g., mass of bodies, ignition power of fuels, etc.), so as to get very basic models (like the gravity law, pressure–volume relationship in a gas, and so forth).

Subsequently, I was trained to elaborate from them more complex model through logical tools, with the support of experimental evidence here and there. In the end of this path, I was able to understand the functioning of the internal combustion engine. Back then, my favorite computational tools were the slide rule, as in the figure below, and an electromechanical calculator (the one belonging to my professor).

Then, exactly toward the end of my studies, early specimens of modern computers appeared in universities and big factories.

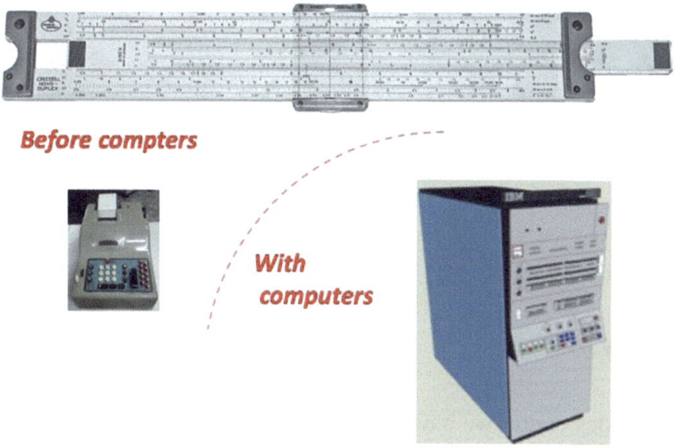

This induced a progressive re-equilibration of the two branches of the data–model loop. Indeed, numerical methods fostered solutions of the reverse engineering problem: "which model may have generated the data we observe?" for instance in relation to combustion pollutants. This evolution may be sketched as follows:

With the availability of computers, inverse problems are not necessarily a curse.

- **Before Computer in Operation (BCO)**: limited inverse problems, huge direct problems (i.e.: infer elementary models from data, then elaborate complex models from them and check their suitability on data).
- **After Computer in Operation (ACO)**: huge inverse problem, slight direct problems (i.e.: infer complex models directly from data, then handle them moderately and check result suitability on data).

For instance, in the latter context, we expect a neural network directly controlling the exhaust emissions of modern car engines. Here the inversion problem coincides with the training of the neural network to learn the function which regulates those emissions depending on various operational parameters of both engine and driver.

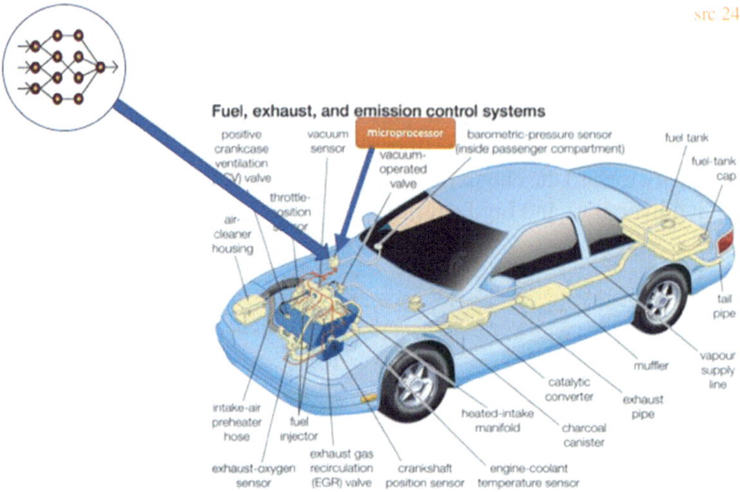

4.6 Memory

Connection weights and biases: the place of long-term memory in a neural network. Frozen weights and biases: the place of rules.

Most of the physics matter we have discussed up till now involves neural networks with fixed parameters, whereas the most relevant activities of our brain refer to the adaptation of its synapses, an attitude we generally denote as learning. The result of learning is in turn related to memory, as a repository of what we have learned. No any hazard will be on my part to enter the complex mechanism ruling learning in the humans. I simply refer to artificial neural networks and standard learning algorithms such as back-propagation. In this framework memory resides in the neuron biases and connection weights.

Actually, relating linear transformation to the cooperation of neurons we may say that with fixed parameters we may model the cinematic laws of a particle in a plane as an iterated application of a linear transform to the six-dimensional vector $\mathbf{v} = \{a_1, a_2, v_1, v_2, s_1, s_2\}$. For initial condition: $(s_1, s_2) = (0, 0)$ (where s_i is the value of the i-th coordinate), $(v_1, v_2) = (0, 0)$ (where v_i is the value of the velocity along the i-th coordinate), and constant values of the accelerations (a_1, a_2) along the coordinates, they read in matrix notation:

$$\mathbf{v} = A\mathbf{v} + \mathbf{b}$$

with

$$A = \begin{pmatrix} 1 & 0 & 0 & 0 & 0 & 0 \\ 0 & 1 & 0 & 0 & 0 & 0 \\ 0 & 0 & 1 & 0 & 0 & 0 \\ 0 & 0 & 0 & 1 & 0 & 0 \\ 0 & 0 & 0 & 0 & 1 & 0 \\ 0 & 0 & 0 & 0 & 0 & 1 \end{pmatrix}; \quad \mathbf{b} = \begin{pmatrix} 0 \\ 0 \\ a_1 \\ a_2 \\ v_1 \\ v_2 \end{pmatrix} dt$$

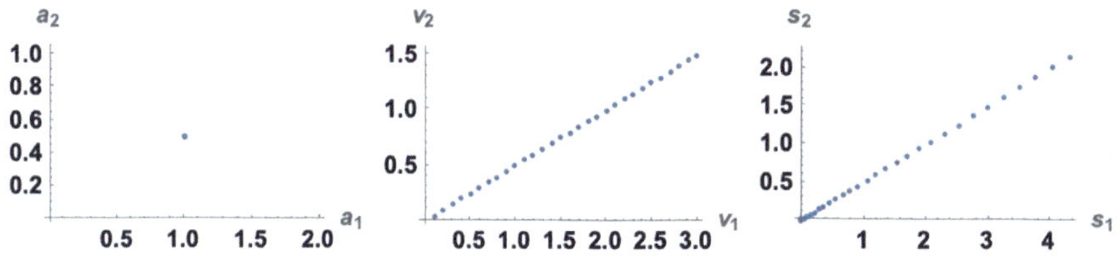

Like in any dynamic equation, the state vector **v** represents the short memory of the system, telling us from which point the next move starts. It's a kind of obliged memory to proceed along any path, so that it is considered an empty memory in the random processes. This is well illustrated by the example of the lifetime of a light bulb. Its exponential distribution says that the probability of the bulb still working after a time t, hence that its lifetime T is greater than t after it has worked already for a time t_0 is:

Current status: the place of short-term memory.

$$P_0(T > t) = 1 - F_{T|T>t_0}(t) = e^{-\lambda(t-t_0)}$$

hence diminishes with t, since sooner or later it will breaks, but it depends only on the remaining time $(t - t_0)$ no matter how long t_0 has been, actually, with no memory of the past. This is a rather ideal condition however; complex systems, living organisms included, are used to maintaining memory of the past, as for both drawbacks such as aging and for benefits such as improvement in skill levels.

Short-term memory however plays a great role if paired with a massive recursiveness. In our brief excursus we have seen that time has been one of the primary quantities considered in physics. Only recently has the focus been shifted to communication, narrowing the border between non living and living entities. The massive implementation of communication leads in many cases to the realization of recursiveness, i.e. of an almost neverending loop where the values of the boiling up variables are updated on the basis of their past values – a kind of deep meditation. Now, if we run infinitely the above system, the various coordinates in the last two graphs go to infinity. This is a typical curse of dynamic systems which requires some wisdom to be avoided.

Short-term memory plus proper communication → recursiveness.

A typical example of a non exploding system that still maintains meaningful features are the fractals. From a pictorial perspective some are wonderful and easy to compute. Moreover, like those in the figure below, they recall shapes of biological organisms in nature. Namely, left picture is computed by me, right picture is implemented in nature, as it represents a bronchial tree in mice lung [20].

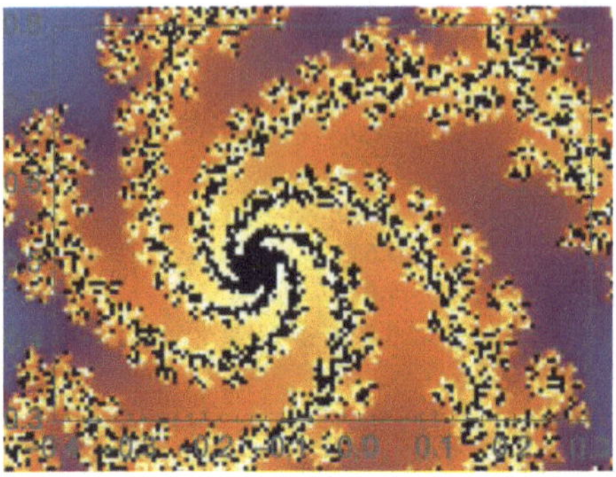

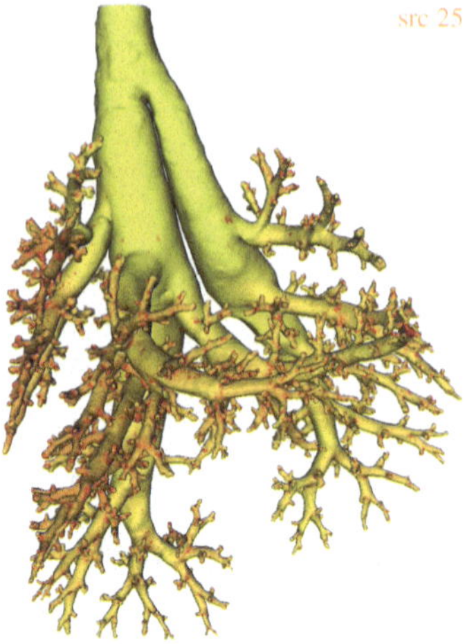

Let's consider the archetypal one, the fern which represents an archetypal plant as well.

What changes with respect to the generation of the above (s_1, s_2) graph is the arrangement of the linear transforms. In the former we have a cascade of linear transform whose task is to uniformly accumulate acceleration effects on velocities and velocity effects on the space coordinates. We can rearrange the transform as a set of the three transforms

$$(a_1, a_2) = \begin{pmatrix} 1 & 0 \\ 0 & 1 \end{pmatrix} .(a_1, a_2) + \begin{pmatrix} 0 \\ 0 \end{pmatrix} \quad (v_1, v_2) = \begin{pmatrix} 1 & 0 \\ 0 & 1 \end{pmatrix} .(v_1, v_2) + \begin{pmatrix} a_1 dt \\ a_2 dt \end{pmatrix}$$

$$(s_1, s_2) = \begin{pmatrix} 1 & 0 \\ 0 & 1 \end{pmatrix} .(s_1, s_2) + \begin{pmatrix} v_1 dt \\ v_2 dt \end{pmatrix}$$

Non exploding recursive system with moderate changes of parameters → fractals;

The fern trajectories synthesize the kinematics of their points through only one linear transform on s_is, where transform matrices are no longer diagonal, no longer with coefficients 0 or 1 along the diagonals. In a certain sense, in place of driving their dynamics from the additive contribution of external variables (velocity from acceleration, space from velocity) they have an inner dynamics of their own ruled by the coefficients of the transform matrix. Moreover, these coefficients may change in a discrete way within a limited number of options. Namely, for the fern instance in the above picture, the options are

$$(s_1, s_2) = \begin{pmatrix} 0 & 0 \\ 0 & 0.16 \end{pmatrix} .(s_1, s_2) + \begin{pmatrix} 0 \\ 0 \end{pmatrix} \quad (s_1, s_2) = \begin{pmatrix} 0.85 & 0.04 \\ -0.04 & 0.85 \end{pmatrix} .(s_1, s_2) + \begin{pmatrix} 0 \\ 0.16 \end{pmatrix}$$

$$(s_1, s_2) = \begin{pmatrix} 0.2 & -0.26 \\ 0.23 & 0.22 \end{pmatrix} .(s_1, s_2) + \begin{pmatrix} 0 \\ 0.16 \end{pmatrix} \quad (s_1, s_2) = \begin{pmatrix} -0.15 & 0.28 \\ 0.63 & 0.24 \end{pmatrix} .(s_1, s_2) + \begin{pmatrix} 0 \\ 0.44 \end{pmatrix}$$

The additive terms induce jumps in the trajectories that are different thanks to the options exerted on the transform in an amount that, in our study case, is proportional to the numbers $p_1 = 0.1$, $p_2 = 0.76$, $p_3 = 0.07$ and $p_4 = 0.07$. A common feature of the above matrices is their contractiveness, in the sense that they map any segment in a segment of shorter length. The rationale of such a kind of transforms, denoted iterated function system (IFS), may be easily captured by the sequence of the other famous fractal, known as the Sierpinski triangle.

It consists of a neverending series of stages, where at each stage the current picture becomes a sub-figure of the one appearing in the next stage. Said in other words, it looks like Nature, once has learned a suitable algorithm, it remembers to apply it at any time the input is available, possibly on a different scale.

A neural network can be even richer as to its behavior: recursiveness is just a matter of stating loops in the network, either a self loop, where an edge starts from non exploding recursiveness plus adaptive parameters may populate a world.

and ends in the same neuron, or a ring connecting the same neuron after a tour through other neurons. In both cases, the involved neurons process the input signal a possibly infinite number of times, like the IFSs. The additional facility is that, during the processing, the connection weights and bias may change if the neuron is involved in a training process.

Actually, training a recurrent neural network is very often a curse, since the weights and biases may either explode by reaching overflowing values or practically collapse to zero. In both cases the network becomes useless. Nevertheless, our brain is a living organism, continuously updating the synapse parameters and rerunning signals in a recursive way. Thus, simulating this attitude is a challenge in view of increasing the results that the neural net paradigm provides us.[16]

4.7 Computation, Communication and Memory: The Platinum Triangle

The Turing machine, as reference model of any computer in the brain of most researchers, contains tools only for algebraic manipulation and memory. Namely, recalling the DTM definition on Sect. 3.6.5,

- the machine Control Unit, corresponding to the CPU of our computer, executes the algebraic manipulations,
- the tape dynamically stores sets of symbols that contain both the primer (the input) of the control unit activity and its output. The latter consists of sets of symbols, representing currently ancillary data of the computation, and the dynamic of the tapes which, through forward and backward moves determines the position at which to store the ancillary data and the position from which to download the new input to the control unit.

Under the agreed notion of non ambiguous function from an input to output, this simple machine is capable of computing any function we may device.[17]

Communication was initially an *orthogonal* tool to share results, like in the left part of the next figure.

[16] Standard computational tool of deep neural networks normally tackles recursiveness only at the level of self loops, where it is relatively simpler to control the mentioned explosion and collapse phenomena.

[17] In turn, we assume that a function is computable if it may be computed by a Turing machine [17].

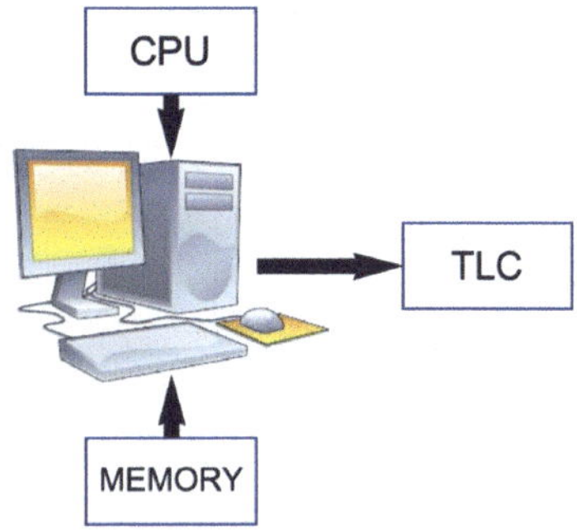

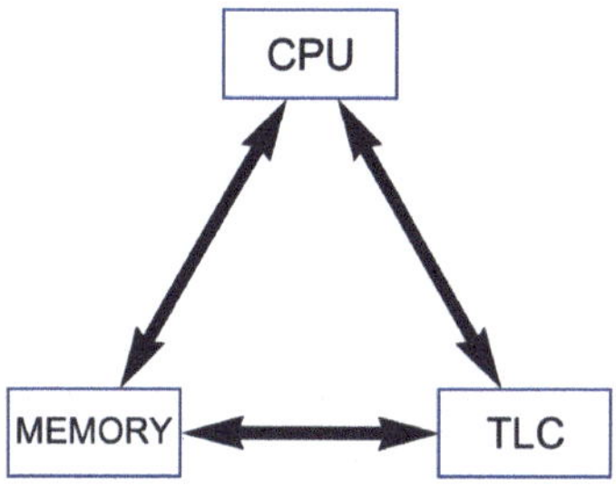

No machine intelligence today without computation, storage and sharing, it's a must of environment conditions,

Nowadays, in the age of cloud computing, computing without communicating is unthinkable. This is not simply because the computation sources are remote (in the cloud), but because the idea of sharing that is inherent in communication attains all the aspects of computation in its current acceptation of artificial brain activity. Thus, the three facilities, computation, communication and memory, began intertwining, like the vertexes of the triangle below. It's a kind of perfect storm, where, like with chicken-egg dilemma, it's difficult to establish which came first:

- communication technology, making the sharing of data and commands easy and almost immediate,
- quantum mechanics, enhancing the dual role of particles as elementary (0/1) data (the particle mode) and elementary communication act (the wave mode),
- the genomic advances, deputing living organisms as the memory of their evolution,
- the cognitive models, declaring memory to be a dynamic synthesis of what the brain has elaborated by,
- feedback exploitation, lending dynamics to computations.

where humans share more complex messages than particles.

Communication puts in share individuals with their own features, where the difference between silence and communication arises if the shared messages can modify these features. With elementary particles, the law of interaction is pre-established, with consequences like entanglement, interference and, in the perspective of many researchers, quantum computations. With more complex actors like animals and humans at the top, things become more intriguing. With complex systems,[18] such as world economy, Earth's climate and even our brain, communication is the glue maintaining the coherence of the systems.

Three main functionalities regulating communications.

First of all, the participant in the communication should agree on the meaning of the exchanged messages, then they must exploit the content of the message and, in the most evolute assemblies, the exploitation should derive from a cooperation between the partners that must be learned in progress. This raises three intertwined functionalities that are at the basis of most aspects of our life:

[18] Complex system is currently a topic supported by a specific discipline. According to [25], a complex system is an ensemble of many elements which are interacting in a disordered way, resulting in robust organisation and memory.

1. clustering
2. optimizing
3. feedback.

4.7.1 *Give a Meaning to Things*

Com-municating what?

Giving a meaning to things, our life included, is a primary goal of our brain. While philosophers of times past preferred tackling this task in solitude, through deep meditation, possibly in a hermitage, modern humans like to find solutions collectively. Apart from the aberrations of influencer and witchdoctors populating social networks, this trend finds its rationale in the awareness that our synapses are driven by the environment, other's synapses included, so that at a given degree individual solutions mate with the ones of the collectivity we belong to.[19] Thus, we may split our meaning giving goal into two intertwining tasks:

1. invent a meaning
2. share a meaning.

4.7.1.1 Naming the Things

No naming without order-ing.

And God made the firmament ... And God called the firmament Heaven. Numerous mythologies emphasize that the creation of things was, at the same time, the naming of things. More modestly, I was used to introducing this task during my lectures by recalling the usual commitment we may have received by our old parent to tidy up the attic of our home, a place where we'd find lots of old things accumulated over the time. Many of these things we never suspected the existence of, others we neither remembered nor imagined any possible use. In any case, tidying up implies that we dispose in some suitable order the items, eventually grouping some of them into shelves. In some sense, we must reinvent the meaning of these objects – like the first man Adam put by God in the world – hence to give a suitable code (name) to each of them. In an abstract sense, the final aim is to substitute the broad bullets we have around the points representing them in a feature space with Boolean functions including groups of them and figuring as their code.

An obvious strategy we may follow in grouping objects consists in similar objects on the same shelf, dissimilar ones on different shelves. These groups of objects will be named *clusters*, understanding that they become *classes* once we acknowledge their utility and reward them for this by giving them a name. For instance, everybody knows what rice is, since it is a suitable and desirable food all over the world. On the contrary, as yet we have no commonly accepted classification of unsuitable e-mails.[20]

Dis-cover-ing pre-existing names,

The clustering problem has been an ideal application field of the decision theory in the Neyman approach sketched on Sect. 4.4.2:

[19] Leave the free will question to priests as an axiom and to physicists as a condition.

[20] This is the reason why the anti-SPAM filters may be different in the mailboxes of different users.

- We start from a hypothesis concerning our shelves. To address a very elementary instance, we may imagine having the set of points $\{y_1, \ldots, y_m\}$ in a two-dimensional feature space, as in the picture below, and that the similarity between two points is decreed by their Euclidean distance (with proper scales of the coordinate axes). In particular, the hypothesis is that these points are distributed between k clusters (three ones in the example below), where the distribution law inside a cluster is a two-dimensional Gaussian distribution law.[21]
- Imagine we already know the parameters θs of these distributions, so that we may lay out the distributions of the clusters on the two-dimensional feature space and visualize them through curves of equal probability density, which are ellipses, aka remarked in the footnote.

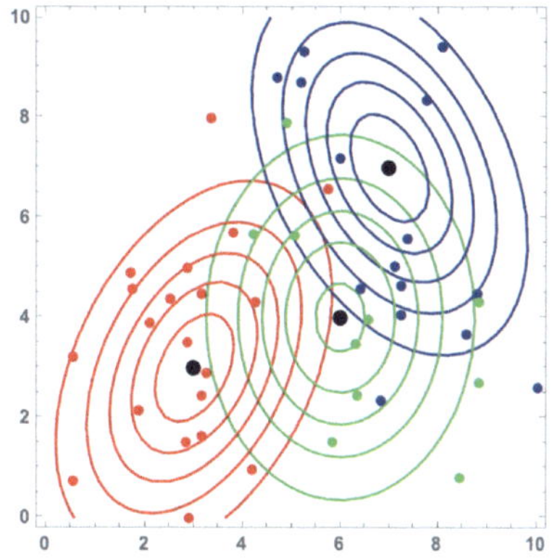

- In the picture above, we realize that there is not a sharp separation between regions where the points of the three distribution may fall. Nevertheless we **must** partition the feature space so as to associate a name to a point, depending on the region where it falls.[22]

---

[21] The shape of the probability density function of a one-dimensional distribution law is the Gauss bell as in the left picture of the figure below. The shape of a two-dimensional one is a three-dimensional bell as in the right picture of the figure possibly distorted such that, if we cut the bell horizontally, we obtain elliptical sections whose axes remain the same when the cut height changes.

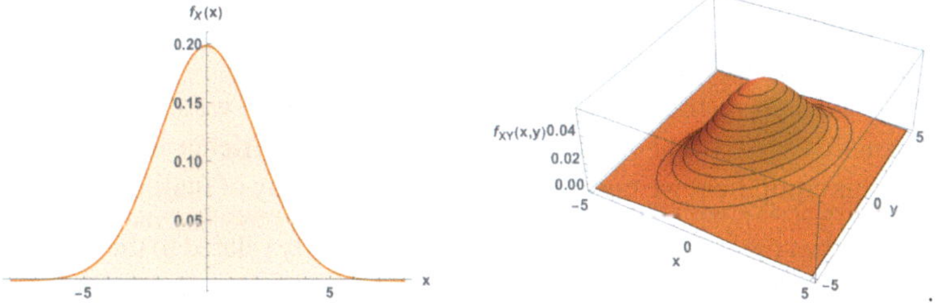

[22] People made wars to establish these rules on their territories.

- Therefore, aka with ethnic minorities, we establish that the partition rule will entail the misnaming of some points and we look for the lowest cost of this drawback. This is the classical decision theory. It consists of:

  – Establishing a *loss function* $\lambda[D(y)|d_i]$ that penalizes mistakes, namely the one of assigning y to the cluster with name $D(y)$ when the *true* name of it, i.e. of the distribution which generated it (according to the hypothesis but hidden to us), is d_i. The most elementary is:

by punishing mistakes.

$$\lambda[D(y), d_i] = \begin{cases} 0 & \text{if the true name of } y \text{ is } d_i \\ 1 & \text{otherwise} \end{cases}$$

  – Defining the risk $R[D_i|y]$ by averaging λ with respect to the conditional probabilities $p_{\Lambda|y,\theta}(d_j)$s, for $j \in \{1, \ldots, k\}$ and the overall risk R of the clustering rule by averaging $R[D(y)|y]$ over Y. With the above loss function, $R[D(y)|y]$ is simply proportional to the probability that the assignment on y is wrong, i.e. different from the true name of y, and R is obtained by averaging this risk on Y.
  – Finding a naming rule that minimize the risk R. The solution provided by the decision theory is in two moves:

    1. Minimize R by minimizing $R[D(y)|y]$ individually on each y
    2. Get the above by assigning y to the name whose corresponding value of the probability distribution function on y is maximal.

But who states in advance the names? Better still, let's establish what we are interested in, i.e. the metrics.

The favorite approach of this book is as an *agnostic* one. Again, it's a matter of deciding to which of possible worlds to assign an object. At this stage, we do not know anything of these worlds. The first thing we must decide is a metric through which to compare the objects in question. Thus, if they are dishes in a cooking competition we may be interested on their taste and presentation; these may be the coordinates of our feature space. The metric could be rather intriguing, possibly rooted on the evaluations of a gourmand commission.

Things become simpler if we must cluster the members of a shooter club, who are possibly more skilled than the one on Sect. 2.1. In this case the feature space may be a Cartesian plane where the target lies, features are the shot coordinates and the metric is an Euclidean distance where the coordinates have the same weight. Said in other words, we adopt an Euclidean metric since we have no reason to privilege one coordinate over another. This may be a very common condition in many operational frameworks, and it's why Euclidean distance is so commonly used.

Thus, features and their metric are the identity card of the object we are managing. The more information we have about the object, the complexer/richer features and metric are. With this equipment we may establish a conjugate way of finding clusters.

Assign the item to the most sensitive cluster, in search of

- The loss function has the same definition, but may be directly related to the metric, without any pre-filled hypothesis. For instance, whatever the clusters we identify will be, we may be interested in having points belonging to a given cluster to be very close, according the distance notion corresponding to the metric, to the *center* v of the cluster, for any definition of center that is congruent with the metric. Hence the loss $\ell[D(y), d_i]$ we define is as follows:

$$\ell[D(y), d_i] = \begin{cases} 0 & \text{if } D(y) = d_i \\ d(y, v_{d_i}) & \text{otherwise} \end{cases}$$

Note that now the loss function is a negative indicator of the appropriateness of a given cluster, indeed it measures how much we pay if we move the name of the object from a given $d_i = D(y)$ to any different d_j.

- Now, we focus on a cost $C[D_i|y]$, as the overall cost of attributing y to any cluster different form cluster i and evaluate it as simply the sum of $\ell[D(y), d_j]$ of all $j \neq i$ (since for $j = i$ the loss is null by definition). Accordingly, the overall cost C is the sum of $C[D(y_r)|y_r]$ on all y_r we must clusterize.
- Note that within this framework, where we know nothing about the *truthfulness* of any cluster, our objective is to maximize the overall cost C. That is to say, we look for an assignment function $D(y)$ so that any change on this assignment gives rise to a high damage, even better, to the most damage. Also the solution provided by this approach passes through two moves:

no truth, but benefits, i.e. maximize the damage we could incur with a different assignment.

1. Maximize C by maximizing, $C[D(y)|y]$ individually on each y, like in the picture below,
2. Get the above by assigning y to the name whose shift is most costly.

In the picture below, starting from the distances in orange of the point ♦ from the centers of the three clusters, respectively in red, blue and green, we evaluate $C[D_i|♦]$ for each cluster index, and assign ♦ to the second cluster.

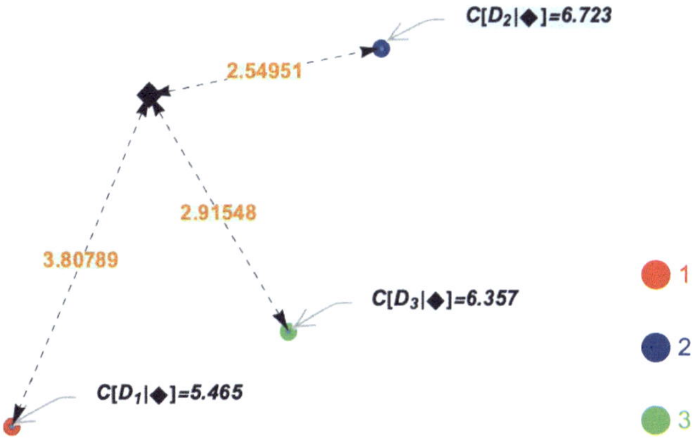

The two approaches are not necessarily divergent. Rather, the second one has the benefit of removing a bulk of superfluity. In fact, in the case that the variances of the three Gaussians in the first approach are equal and the correlations are null, assuming the center of a cluster to have the mean coordinates of the point y_i's assigned to it, the two methods result in the same operational rule: "Assign the point to the cluster with the closest center". By converse, cluster analysis is a vast research filed where many clustering methods have been assessed to "invent" the names of the objects, but not always do the two approaches coincide in their implementation. We mention that the kernel trick is often profited also by clustering algorithms. However substituting the operator "·" with a kernel entails a mapping from the original space to a feature space whose semantics may prove unnatural. For instance, substituting "·" with a Gaussian kernel to compute the distances of points in a *k-means* algorithm[23] may provide a brilliant solution if we want to cluster the red and blue data in the picture below. But this requires to reformulate the algorithm, since the cluster center to be implemented is an average of the clustered points in the feature space and not in the original one. So, we normally make related computations by referring to its definition, rather than a precise value in an infinite dimensional space.

To be handled with care.

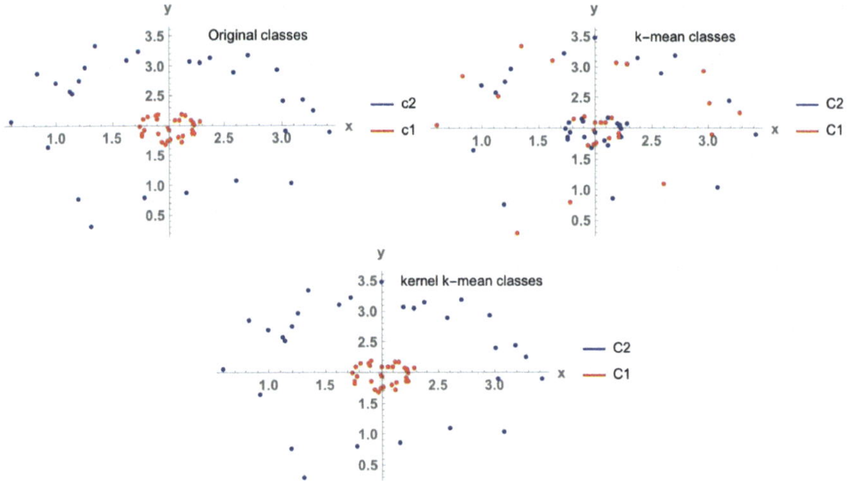

4.7.1.2 The Data Structure-Probability Dualism

As recalled earlier, one of the first conceptual tasks requested in primary school is to synthesize a story, like a Aesop's fable or a favorite movie. The best strategy is to rephrase the story in terms of the essential facts it is composed of and their mutual relations. Rather, if the pupil just repeats the same words as found in the story, then the teacher may guess he was unable to understand their rationale, i.e. the story looks for him like a random sequence of words.

The information lies in the relations between data, i.e. their structure,

Vice versa, in order to understand how the much data along a file are structured, Kolmogorov considers the shortest program π that, in input to a general purpose com-

[23] It is the most popular clustering algorithm which evolves by progressively attributing points to the one within k clusters with closed center. In turn, starting from randomly selected initial points, the centers are the average of the clustered points [3].

puting machine like our laptop, outputs exactly the file. He identifies the complexity $K(x)$ of the file x with the length of π and assumes random those files that cannot be produced by short programs (shorter than the length of x, apart from constants).

The notion of Kolmogorov complexity should be a bit cleaned to be well explained, so that it may emerge its independence, apart for a constant, on the machine where π runs, provided it is a universal machine, and its operating system too, as well as the limited dependence on the input of π.[24] The notion we want to focus on is the structure underlying a string, for instance, of binary data, so that the string $(1, 1, 1, 1, 1, 1, 1, 1, 1, 1, 1, 1)$ has very low complexity (is the string "13 time 1"), while the string $(0, 0, 1, 0, 1, 1, 0, 0, 0, 1, 1, 1, 0)$ needs a longer program to be reproduced.

i.e. in the succinct code of their description,

The second point is that, since the probability distribution law of a variable X is a way of synthesizing the overall population of X, this law must be connected to the above complexity. This is what is captured by the Levin inequality[25]

which we may use in their distribution law,

$$f_X(x) \leq 2^{-(K(x)-c)}$$

Said in other words, we cannot be detailed in the probability density of a variable if there is almost nothing available to describe it. Thus, consider the uniform random variable U made up of n bits. Among its values, no one has a special reason to be observed, hence no code may describe this reason. So, apart from special cases such as the above "all 1 string", the shortest code to compute a string x is the one saying "output exactly x". Hence $K(x)$ is about n, where $f_X(x) = 2^{-n}$.

We may adopt an analogous argument for those variable Y whose relevant metric is the Euclidean one. It says that the sole rationale for distinguishing one y_i value from another y_0 is its Euclidean distance d, which takes on the role of n in the previous example. In other words, for a fixed number n of bits, to move from any point (say $(0, 0, \ldots, 0)$) to any other point described by the same number of bits we need to cover a *coding distance* of around n bits since there are no shortcuts in general. To reach one within the whole set of the points at a distance d, likewise, we must cover a coding distance of around d bits, because distinguishable points are in number order 2^d, for a given granularity, and likewise, no shortcut exists in general. Rather, we refer to d^2 in place of d because of the distance symmetry, so that $K(y)$ is around d^2, and $f_X(x) \propto 2^{-d^2}$.[26] This characterizes $f_X(x)$ as belonging to the family

i.e. in their metrics.

---

[24] We formally define a conditional Kolmogorov complexity as

$$K(x|y) = \min_{\pi \in \mathfrak{P}} \{|\pi| \text{ such that } \mathscr{U}(\pi(y)) = x\},$$

where $\mathscr{U}$ is a universal computing machine (more or less, like your PC) and $\mathfrak{P}$ is the set of all possible programs.

[25] Levin, L.A.: Laws of Information Conservation (Nongrowth) and Aspects of The Foundation of Probability Theory. Problems of Information Transmission 10(3), 206–210 (1974).

[26] Note that an analogous argument holds for the Exponential distribution law, where the main difference is that Y may assume only positive values, and this removes the 2 exponentiation. Moreover,

of the Gaussian distributions, where n characterizes the accuracy of the variables and related distances but doesn't change the rationale of the coding distance.

With Euclidean metrics, the two clustering approaches coincide.

Coming back to clustering, we note that in the case the Euclidean metric is adopted to characterize the differences between items, maximizing $C[D_i|y]$ based on ℓ corresponds to attributing y to the cluster whose distance from the center is minimal. But, as recognized earlier, Euclidean distance is the operational counterpart of the descriptional tool represented by the Gaussian distribution law. Now, if we embed this distribution on the risk R based on the loss function λ, after some algebra deriving from the Bayesian approach to translate the probability of a name given the point into the probability of a point given its distribution associatedto the name (with no a priori information available on the correct naming) we find that minimizing $R[D_i|y]$ corresponds to assigning y to the cluster whose corresponding distribution law denotes the highest $f_Y(y)$. In turn, because of the expression of f_Y, this corresponds, again, to the cluster whose center is closest to y according to the Euclidean distance.[27] That's why, as mentioned earlier, the two approaches lead to the same solution in this case.[28]

4.7.2 *Sharing the Names*

If names do not pre-exist, differences are welcome;

In principle, sharing the names is the prerequisite of any communication between partners. However it's a permanently in progress task as well. From a computational perspective it is a cooperative clustering, where the names are invented as a result of the many clustering algorithms operated by the partners on local datasets and variously intertwined with parallel processing techniques. In the classical approach, it is not an easy task as the goal is to recover suitable labels of items, in spite of the fragmentation of the dataset between partners and the different clustering algorithms adopted by the latter. On the contrary, in the agnostic approach the goal is to converge toward a labeling by exactly taking into account the different experiences (different datasets) and preferences (different algorithms) of the partners (immigrants included). As the goal is to maximize the cost function on Sect. 4.7.1.1, the addition of further ℓ's improves the optimization. Moreover, as no label preexists the data, remaining in the thread of approximate theories, we may accept that the final labeling is not completely the same on the part of each partners. Thus, we may expect that some clusters will substantially coincide for each partner, given the amount of structured data supporting them in each subset, while for marginal clusters we may rely on consensus that is negotiated between partners, even not reached in some cases.

not necessarily they must lead to common names.

things are different if we have more specific features than the sole Euclidean distance to distinguish the points.

[27] With proper rescaling of the axes to take into account the possible distortions of the Gauss bell.

[28] We further remark that in both cases there are some parameters that are left open by the solution. They are scale factor and center positions, which must be estimated along the implementation of the clustering procedure, but this is another task.

4.7.3 Optimizing

A main way to reach a consensus either between partners or simply between our own ideas is the optimization of a cost function. Optimization was the brand of most researches in the eighty of the last century. While identifying the minimum or the maximum of a function has been an analytical goal for many centuries, optimization as the *clou* of the solution of a combinatorial problem comes into play with numerical methods with the widespread of the computer utilities. The general thread of these problems, frequently gathered under the hat of combinatorial optimization, is:

Optimization is a general must;

1. elaborate the problem, for instance a business plan, by injecting the best of our knowledge to simplify the problem and reduce the number of feasible solutions – the special realm of Operational Research,
2. identify an objective function whose maximization leads to the solution of the problem (complementary, a cost function whose minimization has the same effect),
3. numerically optimize the objective function.

The objective function may be relatively elementary, such as a quadratic function in the luckiest of instances,[29] since the more structured parts of the goal have been digested during early elaboration steps. Nevertheless, the numerical optimization remains an open problem for many instances of combinatorial optimization. It represent the last button of many procedures whose success, or at least performance, is entrusted with the effectiveness of this step. Actually, many packages and program libraries offer various alternatives, such as the Nelder-Mead, quasi-Newton and conjugate-gradient algorithms, but not always do they provide satisfactory solutions. Thus combinatorial optimization is the salt of the solution of many problems, but, again, it's not the magic wand. We may simply recall the Knapsack problem mentioned on Sect. 3.4 to realize that there is no combinatorial optimization algorithm that can solve that problem in feasible time and with a guarantee of highly satisfactory results on any input.

like batteries for electrical cars, efficient optimization is the current philosopher's stone;

Relieving this impasse is the core business of quantum computing. In the meanwhile, society stakeholders, mainly big companies, adopt the strategy of crossword puzzle designers (see picture below): decide on a word and then give hints to readers so that they can figure it out by connecting all the hints. Substitute word with world and readers with customers, and there you have it! The solution is one of the various possible worlds that meets the stakeholders' interests and satisfies the customers' induced expectations.

quantum computing looks to be a candidate for such stone.

Pay attention, please. The trade-off between the interests of the stakeholders and those of the customers is not an ethical matter per se to be solved in terms of who are the good guys and who are the bad guys. Obviously, it involves ethical aspects of a just society. However, the solution naturally arises from an optimization process determining the evolution of society, like for natural selection in the evolution of

[29] Because of the convexity of the function.

species. The point is to highlight the actual goal of this optimization and wondering whether and how this goal may be ameliorated.

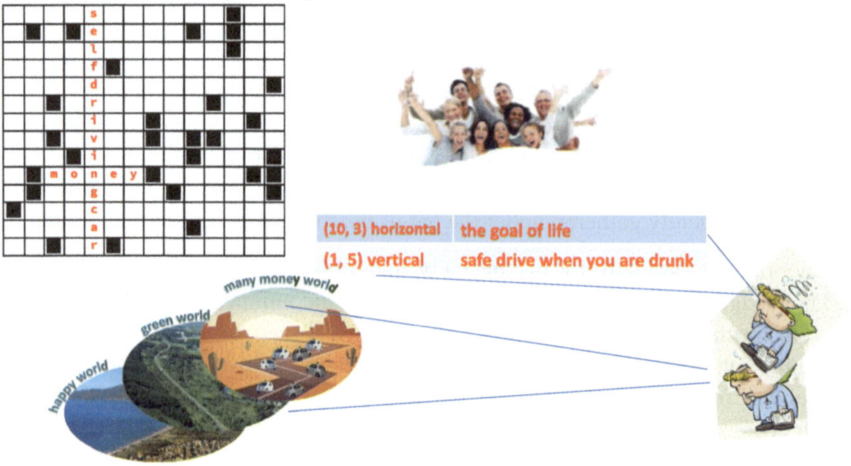

4.7.4 Feedback

Feedback, a tool for either satisfying or cheating the sender.

Of the three ingredients listed above for sharing knowledge, feedback is currently the most claimed one and the most deceived as well. Being formalized by Norman Viener toward the end of the second world war, its concept is very basic. If a baby cries to complain that his food is too hot, we cool it off a bit and serve it to him again. If the baby keep crying we renew the operation until the baby is happy. Sometimes we need rewarm the food, because the previous intervention was carried out too heavy in the opposite direction, but in the end we have the baby to eat the food. With this kind of loops, Viener helped to win the war and triggered the modern system control theory and practice. If we are so skilled to understand the relation between food cooling time and baby crying intensity we may reduce the number of loops, possibly to "0". As quick opposite strategy, if we are a insensitive to their crying an cynical as well, we may force/convince the baby that the food has to be eaten hot as it is. This is the panorama of services 4.0. Any producer of goods or services claims to consider the customer reactions as a precious feedback to his activity. How the feedback is exploited in the span between the two mentioned strategies is generally questionable.

Coming back to our basic neural machine, feedback is an *external* signal through which the machine learns. Thus, we need the machine to communicate with the environment, which in turn reacts to the signal produced by the machine. Then we need an objective function, typically a cost function that interprets the feedback as a fitness measure of the machine in its environment. Finally, an optimization mechanism draws machine parameters, typically connection weights and biases, in the case of a neural network, toward values that maximize the fitness. Since optimization is a hard task, as we realized in a previous section, it is generally carried out through progressive algorithms that are called learning algorithms (when a toddler starts to walk he falls on average 50 times an hour). Independently of the many optimization strategies made available by today's neural network packages, any of us who work in this field has experienced the frustration of seeing the parameters of our neural network either diverging or converging to 0. This occurs even more frequently when the feedback is activated also inside the network (recurrent neural networks). On the one hand these drawbacks make vivid and tangible the physics' statement that the Universe entropy is increasing per se. If not well governed our neurons would lead to chaotic behavior, the antidote to which would be either a continuous training or a scleoritzation as soon as their behavior becomes acceptable. On the other hand we realize the necessity for an analogous need of feedback exploitation, i.e. of learning algorithms also in the *metanetwork* we sketched on Sect. 4.2, namely a network of (brain) neural networks, as a salt of any complex system which perform functions in basically random environments.[30] Currently, this is not a novelty, as any device or service is now fostered by the metanetwork claims to be designed with the help of machine learning intelligence and promises to exploit it during its functioning as well. The question is: with which tools and which goals.

Feedback management at various networking levels.

Actually, the Watt regulator in the picture below is a clean and wonderful example of a quick learning system. The centrifugal forces generated on the two balls by the vertical axis rotation raises their height. In turn this height reduces the steam moving the axis rotation. After some oscillation the two contrasting effects – rotation increase entails ball height increase and height increase entails rotation decrease – find an equilibrium at a certain angular speed. The learning goal is simple: find an equilibrium, and the sensitivity to the parameter ball height is clean, even though non-linear. A more difficult task could be to learn the ball weights to achieve a given equilibrium speed.

Mechanical device regulators, relatively simple;

[30] See [38] for a rather exhaustive introduction to these systems.

Human regulators, definitely more intriguing (luckily!): the goals are ambiguous,

At another extreme of our difficulty scale we find the equilibrium of a complex society like ours. Let us characterize the task through the following features.

- **Goal**. In principle, the general goal is the satisfaction of the society members. Per se the definition of a satisfied society is a political task that currently gets more and more confusing. From a systemic perspective, we could ask whether the goal refers to all members and with the same degree or with which graduation model. Then we may ask how to measure this graduation, i.e. with which metric, whereas the simplest solution is represented on the market by the user experience. In turn, the latter is normally measured with elementary statistics, possibly on boolean variables (the likes), accompanied by some rather uninformative questionnaires[31]
- **Feedback signal**. While the Watt regulator feedback is easily interpretable – "the ball heights are too high because the axis angular speed is high, i.e. the steam throttle is too much open" and vice versa – feedback concerning human satisfaction is extremely difficult to interpret. It derives from a mix of physical and psychological factors, where the former concern both the physics of the human body and the one of surrounding physical devices and physical phenomena. For many years I elaborated on a case study represented by the adaptation of household appliances, such as washing machines, to user habits in compliance with green goals. The blue sheet in the first figure of Sect. 4.1 is an example of a load of laundry requested of the appliance, whereas the satisfaction of the user in the left part of the picture definitely depends on the user's mood. We can try to relate it to some parameters such as softness or cleanness of the washed laundry, or to green perception on the part of the user. However, the function relating the degree of satisfaction to the physical parameters of the wash, such as drum speed, water temperature, kind of detergent, etc., are only partially guessable, while a great part is hidden in

the feedback is generally subsymbolic,

[31] Typical questions: would you suggest our service to a friend?; which competitor's services do you have in mind?

the fuzziness. Thus, the feedback signal is exactly the satisfaction degree, but its exploitation is demanding.

- **The inference engine**. In the case of the Watt regulator the inference engine is a mechanical one. It is represented by the leverage chain which links the opening degree of the steam throttle to the height of the rotating balls. In the study case, as we mentioned before, there is not a very clear cause-effect function. Thus, we must learn this function. Since it is a complex one we may demand the task from a neural network. Due to the fuzziness characterizing the function, as for both feedback value (i.e. the laundry came softy enough) and a reasonable reaction (i.e. to increase softness reduce the drum speed a bit), we may rely on neuro-fuzzy systems like the one in the fourth square of the picture on Sect. 4.4.3. While not delving into the functioning of these learning machines, I underline that these engines as swell need fuel to run, and in this case too fuel is a scarce and costly resource. This is because the feedback comes from human users after a rather longlasting operation (washing cycle) and in conjunction with a wide variety of parameters (laundry specifications and environmental conditions), as sketched below.

the inference engine is complex and slow;

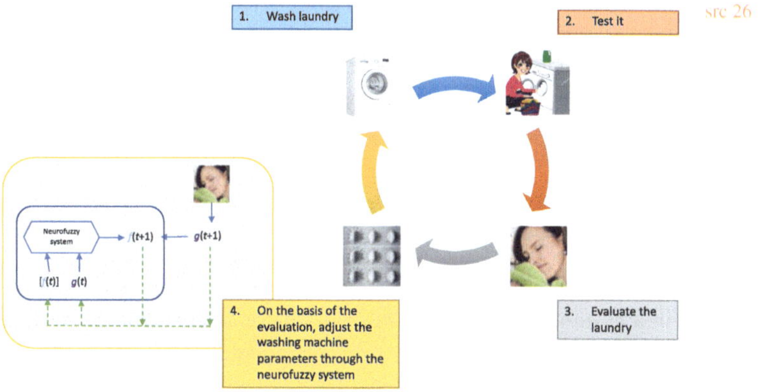

- **Holistic versus distributed solutions**. The handling of the feedback to improve the user experience is a must of any company operating on the market. Handling appropriateness is a question mark whose solution is linked to the goals and to the collection of the feedback signals. Whatever the options, this handling is a social activity made possible on a large scale by the WEB. We may distinguish an *autocratic* approach from a *pluralistic* one.
  - With the former, the scheme is: "I'm a manufacturer who sells his (either material or immaterial) goods to the customer in search of the maximum revenue (it's a declared goal supported by a proper religion). To increase the revenue I take into account requested feedback from the customer in a way that increases the margins".
  - With the second scheme, "We are customers asking a company to produce goods that meet our needs and the manufacturer's need to make a profit".

All drawbacks fostering autocratic solutions.

In the autocratic approach the sole stakeholder is the manufacturer who takes strategic decisions by himself, on the basis of collected feedback. Thus, the related computational center is effectively inside the company, whereas the computing devices, for collecting feedback, elaborating it and making decisions may be indifferently everywhere, so that its location, for instance in the cloud, is merely a question of costs. Actually, also the pluralistic approach could be implemented totally in the cloud. Nevertheless, in order to maintain control over the processing of local information and improving its efficiency, the edge computing paradigm – which brings computation and data storage closer to the sources of data – is gaining ground. This is also due to the need to contain the latency time of cloud computations and other limitations deriving from the central structure, in face of extremely distributed sources and sinks of information and of extremely variegated elaboration needs.[32]

4.8 Further Reading

This chapter has the funny pretense of surveying the wide galaxy of tools for managing uncertain events beyond probabilistic frameworks. Hence further reading should span an ocean of topics and references. But, since I stated right from start that this book contains my own perspective on the matter, I will limit them to a subset of mine which I hope the patient and perseverant reader may find inspiring.

4.8.1 Fuzzy Sets

Fuzzy stets were the result of an operational aim of their inventor, Lodfty Zadey, to manage in a systematic way those graduations on events that derive from experience [42]. After some initial rejection by the Academy, his idea was suited and developed within two parallel veins:

Logics often too complicated;

1. by researchers in logics who found in fuzzy sets an alliance between abstract thought, as a founding tool, and its concrete exploitation in the practitioners' world. However, the common result is a huge list of formal statements to establish the correctness of the fuzzy logic and of the formal tools for its exploitation. All this generally proves arduous to practitioners, me included.

practitioner often too simplistic.

2. by practitioners (including engineers and researchers on data analytics) who found a theoretical counterpart to the experience. The subtle balance to be maintained is between theoretical and practical aspects. Very useful in this respect is the book by Pedrycz and Gomide [33]. I found interesting insight on neurofuzzy systems also in the book by Nauck [29].

[32] Internet of Things (IoT) is the typical example of this framework.

Beyond these two bibliographical suggestions there are plenty of valuable books that the reader may discover by himself according to his own preferences.

4.8.2 Communication

In principle, two brains work differently because there are no synapses from the neurons of one brain the ones of the other. However, if you cut one brain into two you do not get two operating brains. Mostly, this occurs because you break up the complex organization of neurons in a brain. We must keep this in mind when we consider the opposite action of grouping brains. This is a primordial attitude of human societies that lies at the basis of progress. In a very rough synopsis, surrogates of the missing inter-brain synapses have, incrementally over time, become: sub-symbolic messages like gestures, spoken words, written words, and again sub-symbolic messages like biosignals. Actually, sub-symbolic signals regained prominence with the evolution of communication channels that, again from remote to recent times, have been: sensory channels like eyes and ears, printed matter, radio-telephone, e-mail and social networks. Science fiction prophetically anticipated an enormously powerful surrogate as to both content and channel in the form of telepathy (Isaac Asimov's Second Foundation [30]). The underlying assumption is that brains become able to communicate deeply with one another so that the organization of each neural circuit is preserved and virtual synapses feed all signals needed for a powerful integration in an overall neural network, thus creating a *Collective Brain*. Rather than at a hardware level (the synapses) in the graph on Sect. 4.2 we have directly considered the links between local statistics, so evidencing where probabilistic models give way to sub-symbolically drawn models. The human brain is normally well connected in all its parts, apart from those suffering from psychiatric disturbances (like bipolar ones), so that a person's experience consists mainly of statistics within probabilistic models (in turn representing the amount of his coherence). The same does not hold for social networks, where the de-facto: (1) universality of the members and (2) simplicity of their interactions lead far from those models.

Collective Brain versus Social Networks.

Social networks are such common realities that we consider them as a natural environment we prefer to enjoy rather than simply read about. Nevertheless, books like these [37, 41], though notably politically oriented, could help us to focus on some aspects.

4.8.3 Cognitive Algorithms

Cognitive algorithms are a container in continuous expansion. When I started doing research on neural networks, at their revival time around 35 years ago, Neural Networks were a matter for computer scientists, while Cognitive Science was a matter for psychologists and philosophers. A third vein was Artificial Intelligence which at the time was going renewing its identity. Neuroscientists were a distant reference through whom to make an academic connection. Then, a continual convergence process led the three separate areas to merge together while gradually changing their names, and embracing satellites such as fuzzy sets, clustering, and so on. Today the general framework of all this is called "Machine Learning". This process is witnessed by stories of scientific journals. For instance, in 2011, IEEE Transactions on Neural Networks changed in IEEE Transactions on Neural Networks and Learning Systems. Also, new journals, such as Granular Computing, arose to specialize the research threads. A vein than maintained a rather separate course, mainly in the realm of control system research, is the "Reinforcement Learning". The reader may find an exhaustive introduction to the latter in the book [36]

- partly research topic, -partly operational tools.

At the same time, suites of machine learning algorithms have been provided by major companies at the level both of platforms, such as Tensorflow, and API, such as Keras, and of dedicated chips, such as Gaudi and Goya chips produced by Habana.[33]

4.8.4 Quantum Computing

Quantum computing represents a dual challenge. One is to fully understand how to use Qbits to carry out a meaningful computation; another is how to realize a bench of non disregardable quantum bits capable of exploiting the mutual conditioning (the entanglement at the basis of actual quantum computations). As those computations involve code designers, the first challenge split up in two, in its own turn:

A challenge that tries to become a reality.

1. growing developers able to understand the physical phenomena at the basis of their computations. Current developers do not feel comfortable using mathematics and physics in general, at least up to my experience. Moreover, coding on Qbits is hard per se. This is due to the *reversibility* constraint on the operators on Qbits deriving from the physics of the particle interactions. We find somehow analogous "stiffness" induced by the Turing machine model in the classical computations. However, in the latter framework, scientists are able to work at a higher level where they may refer to more abstract and flexible operators coded on our PCs, which hide the complexity of the routines which implement them in the original Turing machine model (like when we say "steer the wheels of a car" omitting to specify the entire list of involved rotations/translations ranging from the steering wheel to the steering gear, up to the drag-link arm and ancillary parts). They work at a higher level too with object oriented programming. Conversely, we are far from similar abstract programming environments in quantum computing. So the second target of the split is
2. relying on a suite of computational tools allowing us to solve real problems.

We may appreciate the matter of these two challenges by reading a book written by the former head of quantum computing in IBM [35] and a recent paper coming

[33] https://habana.ai.

from the same environment [28]. Moreover, Qiskit.org[34] provides an open source software development kit (SDK) for working with OpenQASM and IBM Q quantum processors. The kit offers 11 basic algorithms representing the state of the art and some examples of their exploitation.[35]

As a matter of fact, we appreciate an exponential speedup of the computational time only on some special problems such as the ones concerning particular properties of some functions ($f(x) = f(x+s)$ [34] or $f(x) = x \cdot s$ [16]), for some x and a fixed s.

Nevertheless, a more general purpose one, which extracts records from a file [22], achieves only a quadratic speedup. This is because, like with defuzzification, when working with fuzzy sets, we need to come back from the quantum bit environment where the 2^n variants of n-bits strings may be processed in one shot, and go to the operational environment where the corresponding 2^n results must be suitably reduced to a single one representing the solution of the problem. Thus, with reference to the complexity classes introduced on Sect. 3.6.5, the complexity class BQP, which is the class of problems solvable in polynomial time by a quantum computer with bounded error probability, is generally assumed not to contain the class NP.

As for the hardware, the current number of qubits in operation is a few dozen.[36] Whether or not an analogous Moore's law, which forecasts the doubling of the processing elements every 18 months, will apply also to this technology in spite of the complexity of the environment conditions, and whether or not a wide availability of this computational resource will introduce a second revolution as to achieving numerical results, like the regular computer did in the last century, are still question marks. These questions remain particularly open for people bordering the lion territory with their computations. A general feeling is that quantum computations look more like one shot experiments to be embedded into an overall data processing flow managed by *regular* computers, rather than bricks of new autonomous computing machine. That's why IBM has introduced the notion of *quantum volume* to measure the real capacity of quantum computers in terms of sizes of the *codes* that can be run successfully by these devices, in spite of systematic errors and loss of coherence between qubits affecting the process. This is also the rationale behind the well-known statement by the highly reputed analyst Smith-Goodson[37]: "It means that there will be a lot more things that we can do with quantum computing that we thought were further out on the horizon."

By default, any scientist roots for continual progress in this field. That progress, rather than diminishing their operational field, will empower the machine learning applications (see for instance [1]).

---

[34] https://qiskit.org/.

[35] Not only IBM is working to realize quantum computing devices; for instance, among the others Honeywell is another big company involved in this task.

[36] https://www.ibm.com/quantum-computing/systems/, https://www.honeywell.com/us/en/company/quantum.

[37] https://www.linkedin.com/in/paul-smith-goodson-b6276bb/.

4.8.5 Intelligent Memory

A crucial ingredient of any computation, frequently considered as ancillary matter.

Memory is a complementary attitude of our mind. Sometimes it can play a competitive role: "If you recite from memory a sentence, it means that you didn't understand its meaning." But generally we memorize a sequence of data (texts, movies, theorems, etc.) efficiently because we learned their deep meaning. So, a first demo of the new artificial neural paradigm was rooted exactly on associative memories, whose target is to recover an entire image from a possibly corrupted set of it (for instance reconstruct in our brain the face of a parent from a fragment of an old photo of him) [27].

Speaking of computational complexity we focus in general on run time consumption, whereas what hinders our personal information technology is very often space consumption (think of the stunt we perform almost everyday to find space on our mobile phone to store the last picture we received). Thus data compression became a fundamental functionality of our devices, where a site of compressing tools is customarily available. But we compress significantly because we capture a underlying structure of the data, and this leads us back to the notion of Kolmogorov complexity. An exhaustive book I might suggest on this topic is [26].

A counterpart of the above concept in terms of processes that proves very intriguing to me are the fractals. Citing Barnsley [14] as a reference book is by default. Time ago I made a first attempt of framing fractals in the Algorithmic Inference approach in the paper [12]

4.8.6 Large Numbers in Nature

Large numbers are curse, when they refer to the dimensions of challenging problems,

The size of a problem is a key parameter of its complexity. This has been clear to mankind since the beginning of time. A milestone in human civilization consists of the moment when men began reckoning objects, using as the first means of reckoning the fingers of their hands. Not so strange to imagine, it seems that improvement in reckoning methods was linked to the taxation system, hence to the desire of kings to increase revenues from their subjects. With computers things became easier for computations, for revenue collection too.[38] Nonetheless, big numbers are a critical issue; when they refer to the computational complexity of a hard problem, they represent a nightmare. For example, in the picture below we may see the exponential growth of the running time in seconds which is necessary to solve exactly a knapsack problem (recall its definition on Sect. 3.4) whose instance consists of n items (the *size* reported on the x axis) that are randomly generated, hence expected to escape the worst cases. With this trend we may expect one year of computation when the dimension of the problem grows by a factor 8 (e.g. 640, 000). Vice-versa the sum

[38] Please, don't take this statement as a position on the part of the author against tax collection in general; the revenue system is at the basis of any fair society.

of 640, 000 numbers, for instance belonging to the interval (1–640,000) takes a few hundredths of a second.

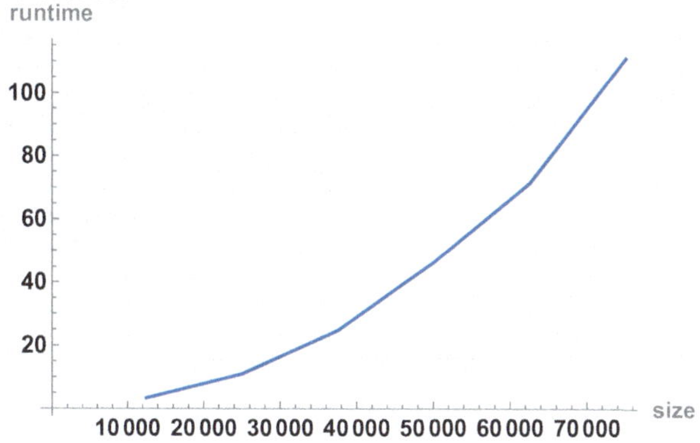

Hence numbers may be large or small, depending on what we manage with them. Our body contains only 20 fingers but order of 10^{14} cells and around as many bacteria. We also recall that the number of protons in the Universe is assumed to be in the order of 10^{78} (the Eddington number), hence around the sixth power of the cells in the body. Hypothetically, if each cell in our body splits in two and each first part of the pair generates a child with all second parts of the cell halves, we would get order of 10^{28} children. So that in three generations we could reach the Eddington number. A somehow similar operation is performed at a different scale by our body reproductive cells. Namely, starting from a pair of strings of 23 chromosomes each, respectively inherited by father and the mother (respectively, shown in purple and blue in the picture below), the male produces in the order of 100 million gametes per day each consisting of a single string of 23 chromosomes, in a renewal cycle so that in the order of 300 million of them are available in any fertilization. Generally, only one fertilizes an oocyte to generate a child, while the gametes not used in matings are degraded. The processing used in this production is such that we expect each new string to derive from a binary assignment of each one of its 23 cells to the homologous chromosome of either mother's or father's string. Thus we expect to find in the overall gametes production all the 2^{23} combinations of homologous chromosomes with a probability that, if no special conditioning occurs in each assignment, is uniform over the gametes, according to the *independent assortment* principle[39] with some caveats [19]. The chromosome choreography is the same for female gametogenesis, but with some differences.

yet a blessing when they refer to the number of elementary processors available to solve them.

[39] https://www.nature.com/scitable/definition/principle-of-independent-assortment-law-of-independent-302/.

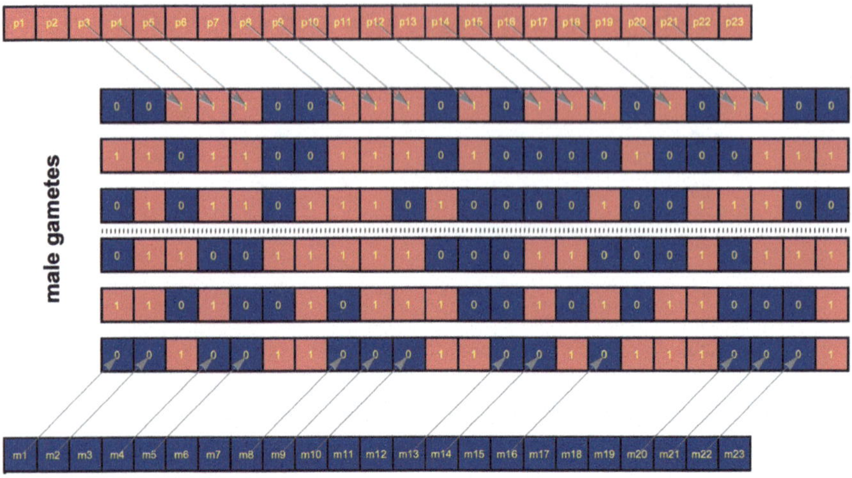

Actually things are more complex, due to the peculiarities of the underlying biological process, denoted as *meiosis*, so that the gamete varieties are even a lot more. While we skip the others, since they are out of our scope, we mention that during the meiosis some genes may mutate according to the local rules mentioned on Sect. 1.4.2.1, where these changes represent a potential enrichment of the overall genetic pool of species. Its actual update occurs by natural selection because only partners who survived till the reproduction age may contribute.

Though not enormous, 2^{23} is a high number that may evoke strategies for biological computers, where the key feature is the parallel computation carried out by plenty of mobile particles (molecules, enzymes, etc.), in number in the order of 10^{14} in [2], driven by nanostructures, in turn made up of macromolecules as well [32].

biological versus electronic processing elements,

Bio-computers and quantum computers represent two complementary challenges to beat the computational complexity barrier currently opposed by combinatorial problems arising in some scientific fields such as genomics, biochemistry and large-scale optimization. The former are based on individual computations performed by an actually large number of processors with a moderate size input, the above mobile particles; the latter on the joint computation of a small number of processors, in the order of the quantum bits, with a huge size input represented by the superposition of all the combination of qbit states.[40] Provided that both alternatives are still at a stage of early prototypes for carrying out specific computations, quantum computing promises enormously more powerful computations, bio-computers currently looks more handy.

The human body is definitely slower than a bio-computer as to the genome replication. While the simple duplication of the genome via the splitting of the host cell, the mitosis, takes dozens of minutes, gametes generation (which occurs in the number of four for each sexual cell) takes dozens of weeks and the virus replication a dozen hours. Rather, our body's computing champion is always the brain: a lot of computing elements, in the order of one tenth trillion, with a clock of around

[40] In number of 2^n, where n is the number of the qbits.

one computation each 5 seconds (an average firing rate equal to 0.16). In the graph below I put the order of magnitude of the pair (number of Processing Elements – PE) and time in seconds of related elementary computations for the mentioned families of computing ensembles. Namely, the values are in Log-scale, *neurons* refer to those in the brain, *computers* refer to a GPU board[41] with 10, 000 cores, *quantum computer* refers to a set of 32 Qbits, which we conventionally split into 2^{32} cores. Since they are based on similar nano-scale technologies, I conventionally attributed the same FLOPS (FLoating point Operations Per Second) to computers, quantum computers and bio-computers (*cells*). From these numbers it emerges our brain to be a well-equipped computing device. Actually, though the duration of the elementary computation is enormously higher than in the case of electronic devices, it proves to be more efficient than they are in performing many tasks. It also suggests that the empowering way pass exactly through a speeding up of elementary computation, a way that is exactly pursued by using computers as prosthetics in the role of a help, but not a replacement, to our brain.

a current competition where FLOPS and feasibility play key roles.

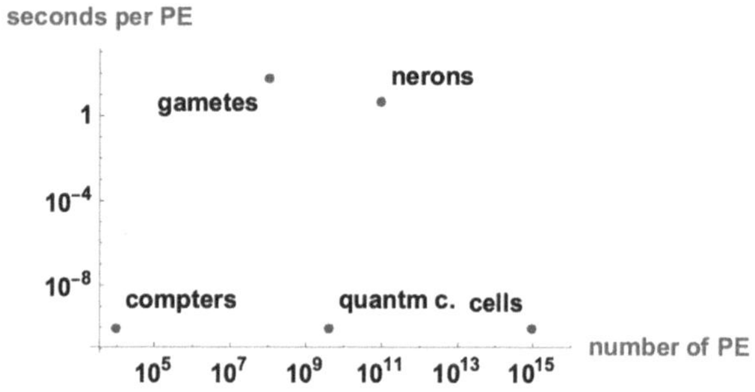

4.8.7 Who Needs Names?

While a large effort on the part of scientists has been made to devise algorithms leading to consensus on names [23, 40] so as to build sharable argumentations, a recent vein has overridden them in view of improving user experience even in the absence of consensus. Said in other words, why would we need to define the names of the conditions that make us happy if in any case we feel happy? In more scientific terms, the goal of identifying the parameters of the clusters according to which we classify items of a population is replaced by the one of training a system to produce items that are very similar to those belonging to a given cluster. Success lies in rendering the user unable to beat the Turing test,[42] i.e. to distinguish whether the next data we meet are generated by the original phenomenon or by a trained machine,

No names, no understanding,

[41] Hosting a grid of processing elements (cores) specially suited to implement matricial computations.

[42] On the one side of a wall, two devices are delivering original and simulated data respectively. On the other side of the wall an algorithm must be able to recognize whether the received value has been issued by the first or the second device.

with the modern additional value that nobody cares to know what the machine is computing. Needless to say, the trained machine is generally a neural network, hence a device which computes non symbolic (hence, normally non identifiable) functions. The two popular training strategies are Variational Autoencoder [24] and Generative Adversarial Networks (GAN) [21].

but powerful simulation tools may replace them, however.

The good face is the enabling of quick and accurate simulations that, on the contrary will help to name the items better. In this sense, the generative model works like an oracle, which provides an as-large-as-we-want set of examples which we may use to identify accurate names for them. The additional benefit with respect to regular bootstrap methods is the rich suite of parameters (for instance of a neural network) that we may train to get a highly reliable oracle.

The bad face is that in this way, for example, we will no longer distinguish whether the responder at the other end of the channel is a person or a machine. Still better, we will no longer distinguish whether or not the environment to which the responder refers is real or is machine generated (the fake news world). *This is a new release of multiverses: not necessarily compatible with the observed data, but compliant with the end user's feeling.*

References

1. G. Acampora, A. Vitiell, Implementing evolutionary optimization on actual quantum processors. Inf. Sci. **575**, 542–562 (2021)
2. L.M. Adleman, Molecular computation of solutions to combinatorial problems. Science **266**(5187), 1021–1024 (1994)
3. C.C. Aggarwal, C.K. Reddyin (Ed.) *Data Clustering: Algorithms and Applications*. ed. by C.C. Aggarwal, C.K. Reddy (CRC Press, 2014)
4. Robert Andrews, Joachim Diederich, Alan B. Tickle, Survey and critique of techniques for extracting rules from trained artificial neural networks. Knowl.-Based Syst. **8**(6), 373–389 (1995)
5. B. Apolloni, G. Avanzini, N. Cesa-Bianci, G. Ronchini. Diagnosis of epilepsy via backpropagation, in *Proceedings of the International Joint Conference on Neural Networks* (Lawrence Associate, 1989), pp. 571–574
6. B. Apolloni, S. Bassis, J. Rota, G.L. Galliani, M. Gioia, L. Ferrari, A neurofuzzy algorithm for learning from complex granules. Granul. Comput. **1**(4), 225–246 (2016). (Dec)
7. B. Apolloni, S. Bassis, L. Valerio, Training a network of mobile neurons, in *The 2011 International Joint Conference on Neural Networks (IJCNN)* (2011), pp. 1683 –1691
8. B. Apolloni, C. Carvalho, D. de Falco, Quantum stochastic optimization. Stoch. Process. Appl. **33**(2), 233–244 (1989)
9. B. Apolloni, F. Kurfess (eds.), *From Synapses to Rules - Discovering Symbolic Rules from Neural Processed Data* (Kluwer Academic/Plenum Publishers, New York, 2002)
10. B. Apolloni, Inferring statistical trends of the covid19 pandemic from current data where probability meets fuzziness. Inf. Sci. **574**, 333–348 (2021)
11. B. Apolloni, A. Piccolboni, E. Sozio, A hybrid symbolic subsymbolic controller for complex dynamical systems. Neurocomputing **37**, 127–163 (2001)
12. B. Apolloni, S. Bassis. Identifying elementary iterated systems through algorithmic inference: the cantor set example. Chaos, Solitons Fractals **30**, 19–29 (2006)

13. APOLLONI B., S. Bassis, C. Mesiano, S. Rinaudo, A. Ciccazzo, A. Marotta. Statistical parameter identification of analog integrated circuit reverse models. Journal of Artificial Intelligence and Soft Computing Research (2011)
14. M.F. Barnsley, *Fractals Everywhere* (Academic Press, 1988)
15. G. Bateson, *Steps to an Ecology of Mind: Collected Essays in Anthropology, Psychiatry, Evolution, and Epistemology* (University of Chicago Press, 1972)
16. E. Bernstein, U. Vazirani, Quantum complexity theory, in *STOC '93* (1993)
17. W.S. Brainerd, L.H. Landweber, *Theory of Computation* (Wiley, 1974)
18. T.G. Dietterich, Ensemble methods in machine learning, in *Multiple Classifier Systems* (Springer Berlin Heidelberg, Berlin, Heidelberg, 2000), pp. 1–15
19. A. Franklin, A.W.F. Edwards, D.J. Fairbanks, D.L. Hartl, T. Seidenfeld, *Ending the Mendel-Fisher Controversy* (University of Pittsburgh Press, 2008)
20. G.J.M. Garcia Lapdmouse: a data archive for advancing computational models of inhaled aerosol dosimetry. Journal of Applied Physiology (2020), pp. 307 –308
21. Ian Goodfellow, Jean Pouget-Abadie, Mehdi Mirza, Xu. Bing, David Warde-Farley, Sherjil Ozair, Aaron Courville, Yoshua Bengio, Generative adversarial networks. Commun. ACM **63**(11), 139–144 (2020). (Oct)
22. L.K. Grover, A fast quantum mechanical algorithm for database search, in *Proceedings of the Twenty-Eighth Annual ACM Symposium on Theory of Computing* (Association for Computing Machinery, New York, NY, USA, 1996), pp. 212 –219
23. P. Jha, A. Tiwari, N. Bharill, M. Ratnaparkhe, N. Nagendra, M. Mounika, Scalable incremental fuzzy consensus clustering algorithm for handling big data. Soft Comput. **25**(13), 8703–8719 (2021)
24. M.J. Kusner, B. Paige, J.M. Hernández-Lobato, Grammar variational autoencoder, in *Proceedings of the 34th International Conference on Machine Learning*, volume 70, ed. by D. Precup, Y.W. Teh, *Proceedings of Machine Learning Research* (PMLR, 2017), pp. 1945–1954
25. J. Ladyman, J. Lambert, K. Wiesner, What is a complex system? (2011)
26. M. Li, P. Vitanyi, *An Introduction to Kolmogorov Complexity and its Applications* (Springer, Berlin, 2008)
27. A.N. Michel, J.A. Farrell, Associative memories via artificial neural networks. IEEE Control Syst. Mag. **10**(3), 6–17 (1990)
28. G. Nannicini, An introduction to quantum computing, without the physics (2020)
29. D. Nauck, F. Klawonn, R. Kruse, *Foundations of Neuro-Fuzzy Systems* (Wiley, New York, NY, USA, 1997)
30. P. Nelson, S. Zyglidopoulos, Learning from foundation: Asimov psychohistory and the limits of organization theory. Organization **10**(4), 591–608 (1999)
31. J. Neyman, E.S. Pearson, On the problem of the most efficient tests of statistical hypotheses. Philos. Trans. R. Soc. Lond. Ser. A, Contain. Pap. Math. Phys. Character **231**, 289–337 (1933)
32. D.V.Jr. Nicolau, M. Lard, T. Korten, F.C. van Delft, M. Persson, E. Bengtsson, A. Mansson, S. Diez, H. Linke, D.V. Nicolau, Parallel computation with molecular-motor-propelled agents in nanofabricated networks, in *Proc Natl Acad Sci U S A*, vol. 113 (2016), pp. 2591–2596
33. W. Pedrycz, F. Gomide, *An Introduction to Fuzzy Sets : Analysis and Design* (Bradford, 1998)
34. D.R. Simon, On the power of quantum computation, in *Proceedings 35th Annual Symposium on Foundations of Computer Science* (1994), pp. 116–123
35. R.S. Sutor, *Dancing with Qubits: How qUantum Computing Works and How it can Change the World* (PAKT> Birmimgham, Mumbai, 2019)
36. R.S. Sutton, A.G. Barto, *Reinforcement Learning: An Introduction*, 2nd Edn. (The MIT Press, 2018)
37. J. Taplin, *Move Fast and Break Things* (Brown & Company, Little, 2017)
38. S. Thurner, R.A. Hanel, P. Klimek, *Introduction to the Theory of Complex Systems* (Oxford University Press, 2018)
39. G. Towell, J. Shavlik, Extracting refined rules from knowledge-based neural networks. Mach. Learn. **13**, 71–101 (2004)

40. T.S. Wierzchon, A.M. Klopotek, *Modern Algorithms of Cluster Analysis*, Studies in Big Data Book 34 (Springer, 2018)
41. T. Wu, *The Attention Merchants: The Epic Struggle to Get Inside Our Heads* (Deckle Edge, 2016)
42. L.A. Zadeh, Fuzzy sets. Inf. Control **8**, 338–353 (1965)

Chapter 5
Out-of-the-Box Solutions Versus Answers for Free

Abstract As it has emerged from the previous chapters, the golden kit allows for compatible worlds running both in *monolithic* and in *distributed* mode. Both are based on the platinum triangle computation-communication-memory architecture and on cognitive algorithms. The main distinction concerns the objective function: (i) either in favor of a few stakeholders and, when possible, with benefits to other nodes of the communication network, as a corollary, or (ii) in favor of the latter with appreciative benefits to delegated stakeholders. Without getting into the hard question of the free will of individuals in the network, I will illustrate the benefits and drawbacks of the two optional modes, obviously from my own perspective that is in favor of the latter of the two. The idea, typical of any scientific writer, is that the numerous examples I'll provide will stimulate the neural networks of readers toward my same option. The illustration style would be a bit engaging, as for topics and figures, but funded on the aseptic and rigorous (as much as possible) argumentations of the previous chapters, so as to capture the convictions of the reader rather than earn a simple like. This won't to be an ethical manifest dividing good from evil; rather, according to the cognitive schemes, the hope is that readers will find these arguments compliant with their world and will share them in their communication networks.

Like with the world timeline divided in *B.C.* and A.D., the life of the whole of science and of human society in general may be dated as before and after the birth of the computer (*BCO/ACO*). Actually, this birth was a rather long deal, lasting either centuries or years, depending on our notion of computer. Personally, I experienced this birth on two occasions: a first time when working on my graduate thesis, and a second time during my first year working for a company. Actually, I'm so old that for my first 20 years there was no computer in my life: only an abacus as a joke, a slide rule for scientific calculations and an electro-mechanical calculator for ease. During my thesis work, in 1969, I met the wonderful *P101 Olivetti* desktop computer, shown in the left part of the picture below,[1] which I used as a comfortable extension of the

[1] http://www.curtamania.com/curta/database/brand/olivetti/Olivetti%20Programma%20101/index.html.

B. Apolloni, *A Few Things I Know About Her*, Intelligent Systems Reference Library 219, https://doi.org/10.1007/978-3-030-94379-0_5

slide rule with additional facilities of automated computations and implementations of light logics such as:

- the *for* loop to repeat computations and
- the *if* conditioning to fork some of them.

Then a year later I met the mainframe *IBM 3260*, shown in the right part of the picture below, at the company where I was working. While I used it with a very wide scope, remaining in fields of scientific computations through Fortran codes, the main jobs of the machine in the company concerned management tasks (such as worker attendance register, pay slip featuring and stock inventory). Actually, the bottlenecks of the company data centers at that time consisted of data entry through punch-cards, fed by a division of various dozens of operators, and printout production, where those data were duplicated, split, rearranged and aggregated at the service of the various divisions, for a yearly cost of the sole paper up to three times the rent of the computer itself. These two lines of operation, scientific and management, proceeded in parallel until merging at the end of nineties with the employment of sophisticated statistics as a common tool for both. Pattern recognition, market forecasting and last cognitive algorithms are application instances of those statistics.

Computers and statistics everywhere.

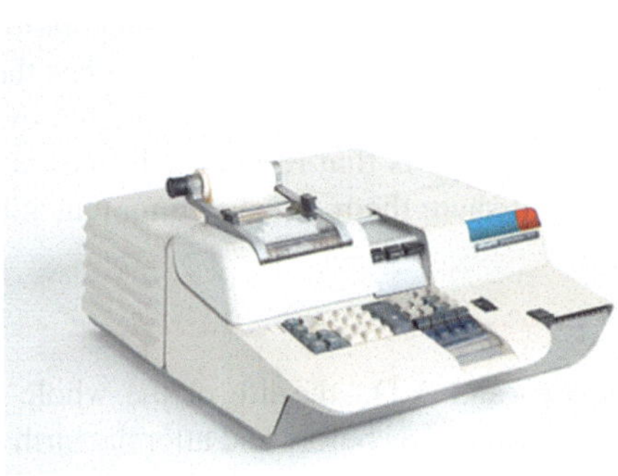

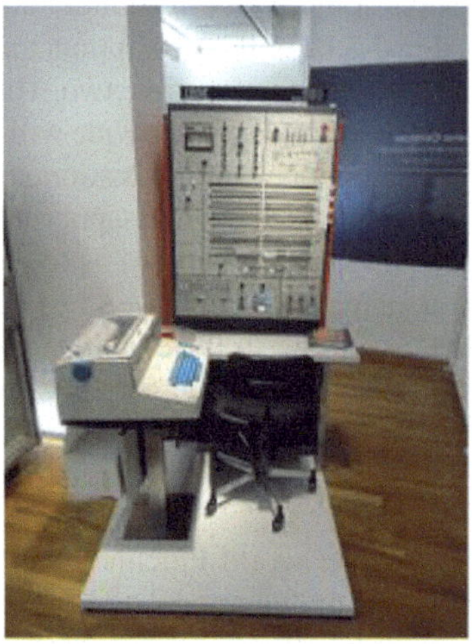

Then comes the entry into play in sequence of two factors: the spread availability of computers and telecommunication networks, which thoroughly changed the ACO era scenarios. With the former anybody can in principle enjoy computing facilities; with the latter just a few people have the power to shape the usage of those facilities. This may revert the great benefit of *touching by hand*, in numbers and graphs, the results of modeling a compatible world into the possible drawback of *deceptively experiencing* a pleasant virtual world which is passed off as the true one.

5.1 The Universal Responder Paradigm

In the current paradigm of the *world as a service*, facing any kind of problem, we are principledly led to ask the web for an answer. Maybe we do so by looking on Wikipedia or Github,[2] when we are inclined to contribute to the solution, or on thematic networks when we just want the solution, possibly selecting from among the ones with most number of likes. Per se this attitude is not new. In antiquity the role of these networks was played by haruspices and diviners, and subsequently by fictitious proto-robots such as Golems as in the Jewish tradition. Recently, the myth of Artificial Intelligence proposed in 1959 by Simon et al.,[3] the General Problem Solver Machine, has come to the fore again as people start asking their *home Universal Responder Machine* (URM) about the solution of complex problems regarding device configurations or stock market investments, for example. The picture below sums up this evolution.

World as a service,

src 27

[4,5]With the introduction of smart speakers like Alexa or Siri, people have begun interacting with general purpose responders that appear to be able to answer any question that may arise in their life. The perception of this facility has been fostered by the seamless coupling of voice recognition with Internet search engines, with a few huge corporations playing the role of Over-The-Top operators (OTT). They feed those engines with questions and deliver the answers, meanwhile harvesting all sorts of valuable information in the process. According to Internetlivestatistics.com,[6] Google currently processes over 40,000 search queries every second on average, which translates to over 3.5 billion searches per day and 1.2 trillion searches per

Universal Responser as any-service provider,

---

[2] https://github.com.

[3] Herbert A. Simon, J. C. Shaw, Allen Newell: The General Problem Solver.

[4] Detail of an Etruscan bronze handle, depicting a haruspex, kept at the Allard Pierson Museum in Amsterdam.

[5] A Prague reproduction of the Golem.

[6] https://www.internetlivestats.com/.

year worldwide. In the period 2018–2019, 26.1 million smart speakers were shipped around the world, with a sky-rocketing outlook which forecasts a market of 35.5 billion US dollars for 2025.

via a unified language

As considered for social networks, URM facility presents both a good face and a bad one. As for the former, we may say that the URM plays the same role that the introduction of the state television broadcasting did in unifying national language use among the people of the relative country. The same is happening today with the specific languages of technology (mainly ICT technology). Note that national unification in terms of a common language did not involve learning about the literature of the country (even though sometimes it represented the primer of higher education). Analogously, receiving from the URM the answer, for example, that for the best TV viewing we need to buy an oled screen, we may learn what to ask the TV vendor for but not necessarily do we know what oled technology is.

and unified thought, as well,

Conversely, the comfort and ease provided by the URM environment leads to a standardization of society through dynamics where the user asks the URM to provide solution to his own problems and the URM delivers answers in line with OTT strategies. Often, these answers do not meet user needs in full; however they are available almost for free and are adopted by the user. This happens not only at a retail level through either explicit or hidden recommender systems (comments such as "Frequently bought together" and "Products related to this item" belong to the former, while enforced ranking of products in search engines belongs to the latter). But also at industrial level companies generally prefer to use standard AI suites to solve their problems rather than get solutions on their own; and, once again, these suites are produced by a few huge corporations. Amazon is a paradigmatic example of this: its business was born as e-commerce, but the success in this venture represented the leverage to become a technology company sharing the main business in cloud computing, digital streaming and artificial intelligence with the other big four in the US (Apple, Google, Facebook and Microsoft). We may cite similar stories in other technologically evolute parts of the world, but the number of the members of this oligopoly remains in the same order.

broadcast through a very limited number of social networks.

We also remark that while any individual endowed with a sufficient technological background may set up a social network (and many thematic networks emerge every day in this thread), the most popular social networks fall within the realm of this oligopoly. It's a rather obligatory pattern, since to preserve its existence the oligopoly must include any initiative that is capable of generating consensus or improving related technologies, still in the thread of standardizing societies and their requests. We see that the entire globe is covered by 5 major social networks, whereas around 20 represent the relevant general purpose networks. To them we may add a large number of thematic networks specializing in a particular activity, hobby, role, etc.

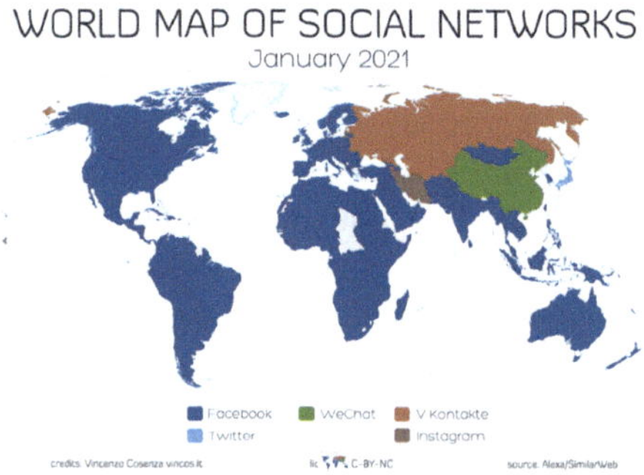

src 28

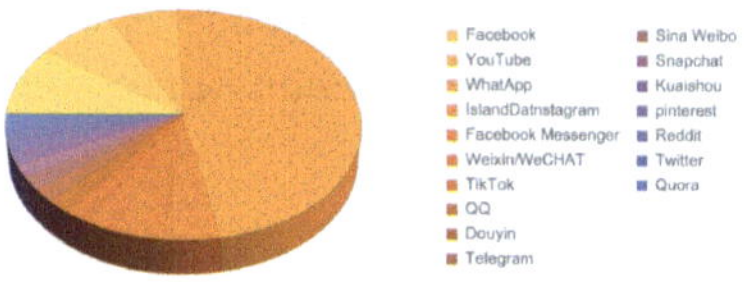

5.1.1 *All That Glitters Is Not Gold*

We revive this ancient proverb in current times to analyze the evolution of the industrial strategy since the eighties of the past century. In 1999 this strategy was effectively described by the Naomy Klein book NoLogo [5], where she explained how the *branding* phenomenon is a way of directing the customer desires exactly toward the products made by the branded company, almost independently on the quality factor, so that the core business of the company moves from the product (now manufactured cheaply by factories anywhere) to the messages through which to promote it. The standardized scenario depicted in the previous section constitutes the continuation of the story. Then the keyword of the latest-generation branding strategy becomes the *user experience*, where the further underlying message is: "don't care how my product is made, provided that I ensure you a wonderful experience of it". It's a rather disquieting message that leads the user environment toward a virtual reality. On the one hand, this may be a waste of the very sophisticated technologies currently put in operation by scientists and may hide concrete benefits generated by them. On the other hand, this standardized world may prove not so effective as to real user experience.

You ask a question, and accept any answer.

Let's consider a few aspects of URM standardization that directly impacts our life:

- query "how to do something." If the question concerns how to make "amatriciana" sauce, I can expect to get different recipes, possibly depending on the tastes of the person whose recipe it is, but this doesn't matter.
  - Things become more severe if I look up info on how to recover some illness of mine. Actually, when we feel bad we are often led to use Google to understand what might be wrong. However, well-reputed medical studies claim that asking the Internet to diagnose a disease based on symptoms produces mistaken diagnoses, which may prove dangerous as well two times out of three. This rate does not take into consideration disastrous answers the user might get upon inquiry from bizarre websites such as witchdoctor pages.

  - Analogous drawbacks, though not lethal ones, we might encounter if we ask for solutions to some software problems. There are plenty websites supplying answers. Many of them provide serious and accurate answers in the true style of open source collaboration. From time to time however we find answers like: "The code lines I suggest should work, but I had no time to test them."
  - things get worse in hacker communities, where hackers are the target of some markets, so that *naive creativity* is encouraged, no matter the correctness of the results.

- query "what is that". "Google Has Picked an Answer for You–Too Bad It's Often Wrong." This assertion, written on the Financial Times in 2017 by Jack Nicas, is the synthesis of many disclaims on Internet Truth (obviously provided by other browsers as well). The incorrect status of the answers may derive from three causes:

  - Mistakes in the sources where the browser pick the answers through a mix of lexical interpretation of the query and reputation of the source. For instance, if we type "scooter sharing Pavia." the answer is a list of renting companies which have nothing to do with the sharing of scooters in the city of Pavia. Forget the "fake news" chapter.
  - Mistakes to favor some products, companies or authors. That's the dark face of free solutions where the sorting of the answers is biased by the above entities. We don't pay the provider directly for the answer, but we pay him indirectly through the conditioning of our choices and the commission by the favored entity.
  - Ambiguity of the inquiry. For instance I had a lengthy quarrel with a friend of mine concerning the political departure of a famous leader in my country. The question was: "Did he retire or was he not reelected?". Since both events are reported, though at different times, depending on the relevance we gave them and our attitude toward him, each one of us claimed to be right.

- reaction "what remains in my own realm". Inquiring on the web about "what Google cannot answer correctly," I found the answer "all that concerns creativity". Well, this appears obvious in the sphere of the humanities, but what about in everyday life? Looking at the rankings of the intellectual jobs most in demand in the tertiary sector, we find medical jobs almost uniformly in top place, for obvious reasons in this pandemic period; then come financial jobs (requiring a mix of creativity and aggressive attitude) and all variants of data analytists (ranging in skills from AI tool adaptation, to network infrastructure and data pre-processing and post-processing expertise). Of course, this is for the mainstream, whereas, luckily, many people try to invent activities with different added value (forget those startups whose sole aim is to be adsorbed by the mainstream at a good price).

5.1.2 *And the Wise Owl Said, Let the Hare Eat the Fox*[7]

Hold on! As a pre-condition of this book, we excluded any discussion of *superior* entities in the category of soul, oracles, divinities and so on. Thus, the URM must be a machine.

Basically, we have discussed two kinds of machine: fixed hardware, upgradable hardware. The template of the latter is an artificial neural network, still better, a deep neural network. We also suggest a way of considering the former kind as a sclerotizastion of the latter so as to provide a unique general architecture.

In any case, a machine with fixed hardware, like our laptop, (i) requires an initial input, for instance some axioms, and queries, then (ii) mechanics of the device enter in operation, and (iii) outputs, possibly in a neverending process, theorems, answers to queries and any desirable data. In essence, it is a logic machine, so that, given the input axioms as a pre-existing truth, it will provide any answer for the future. However, the incompleteness theorem[8] says that we cannot guarantee that it can answer any query. Moreover, since we are relying on a machine, maybe it *could* answer some queries but in a time so long that it becomes useless to our brief life.

A wise owl is never mistaken: operational details are for the inquirer.

Hence we decided to pose these hard queries to a machine with upgradeable hardware. But this solution, too, is not for always. A neural network (any cognitive system in general) learns from examples to answer queries that are similar to the ones on which it has been trained.

And in the case of surprises? A honest answer would come by tossing a coin, and inform the inquirer of this. Rather, the *branded* answer is like the one of the owl gave in the Aesop's fable: an answer that is syntactically correct but does not provide any insight to the inquirer ("Let the hare eat the fox," to solve the problem of the hare's survival). Thus, for instance:

- if a software does not work, "it is not the fault of the operating system, says the responder on the part of the computer company, you must ask the software manufacturer" (and vice versa)
- if the Internet connection is down: "did you try to change the modem?"
- if the vacuum cleaner does not work, "that's very strange, who knows how you used it".

and so on.

---

[7] One of Aesop's fables.

[8] https://plato.stanford.edu/entries/goedel-incompleteness/.

src 29

A third way is to use the logic machine to extend results into new fields and cognitive algorithms to fill up holes with experience. This is what is commonly done in the everyday life; but it may prove too costly and not necessarily strategically convenient for massive usage.

5.2 The Out of Box Solutions Thread

Currently, URM tools are very efficient and rather easy to use. Still, we claim that relying on such tools acritically is leading society closer to the Golem paradigm where any request can be satisfied provided it has been correctly issued to a Universal Responder Machine in the cloud. The Golem paradigm counterpart is the Golem effect, a well-known psychological phenomenon in which lower expectations are placed upon users that rely on external help. This effect has been studied in educational and organizational environments. It is a form of self-fulfilling prophecy. A way to preserve the universality of the URM approach while countering the Golem effect is through adaptation: interacting with the URM, like when using a GPS navigator, we must decide whether to comply in full with the road indications provided by the system or to adopt them only as a suggestion to be re-elaborated personally. This allows us to maintain a diversity of scalable operational solutions with respect to the ones that will be suggested in a nearly monopolistic way by OTTs. Indeed, the sheer size of OTT companies has the effect of leveling out environmental conditions exactly to the ones planned by them, with the beneficial effect that we would be never markedly wrong by adopting their answers. Nevertheless, disruptively innovative solutions may principally arise from out-of-the box thinking.

URM suggestions are OK, but who decides is me,

Multiverses is a keyword of this book, as stated in the introduction. Compatibility is another one, and on the basis of local statistics of my own, the standardized universe is only in part compatible with the expectations of friends and colleagues of mine. So no one at all rejects URM responses.[9] However, many of these people like to maintain a realm of their own where they can analyze those responses and possibly rework and complete them with own considerations, experiences and computational tools. Let me quote a set of veins in this direction calling as a whole for out-of-the-box solutions:

[9] Don't throw out the baby with the dirty water.

- **Production from below initiatives**. The challenge is to setup thematic social networks aimed at fostering activities with a social content in a way that is unbiased by the main stakeholders.

  1. A well known example is creative commons, a sort of app market, where app is to be intended in a loose sense (so that a picture too falls in this market), and the sale is essentially free of charge, possibly with some limitations on the usage. The Linux operating system is a relevant item of this free of charge market.
  2. Another case is represented by the crowd-funding platforms, where operational responses in terms of project's enterprises and/or goods are funded through the collection of small amounts of money among anyone interested.
  3. But convention, political meetings and lectures may also be the object of the production from below initiatives.

with the coop-eration of others at the same level,

- **Explainable Artificial Intelligence**. With this objective, people ask the URM not only for a response but also for the rationale behind it. A paradigmatic instance is Deep Neural Networks (DNNs). With this tool major companies promise to provide solution to very intricate problems. For instance:

  1. Within this family, a convolutional neural network (CNN) has been trained to map raw pixels from a single front-facing camera directly to steering commands. This end-to-end approach proved surprisingly powerful. With minimum training data from humans the system learns to drive in traffic on local roads with or without lane markings and on highways. It also operates in areas with unclear visual guidance such as in parking lots and on unpaved roads. The system automatically learns internal representations of the necessary processing steps such as detecting useful road features with only the human steering angle as the training signal, while it never was explicitly trained it to detect, for example, the outline of roads.
  2. In the early days of 2020, Google's artificial intelligence division DeepMind introduced a deep learning model that reportedly improved results of an average radiologist by 11.5% and significantly reduced the workload of the second reader in the British scenario.
  3. DNNs are applied to predict financial market movement directions. In particular, long short-term memory (LSTM) modules are fed with raw data to forecast these directions.

All these topics exhibit brilliant results generated by deep neural networks as the core of the decision system. But who will entrust his live with the above CNN based driving simply because during the tests it performed well, without any cogent explanation of why? Analogous questions concern the diagnosis of cancer and the fate of personal stock options. Whitening the black box generally represented by DNN and by most of the current AI tools is a goal that is cyclically pursued by the researches over the span of time. In view of the URM scenario, current times demand results for this goal.

Actually, a sharp distinction is made by academia between Explainable AI systems and Interpretable AI systems. With the former we expect an explanation of why the provided I have understood problem and solution.

black box took a given decision as a function of the input and environment, where the explanation is essentially requested to be expressed in natural language so as to be understood by the *lay* inquirer (typically our politicians, both in the government and in other managerial contexts) as a specific form of user experience. Rather, interpreting is a matter of sufficiently cultivated people, so skilled as to be able to understand elementary analytical expressions putting in a clear, though possibly approximate, relation the input and output of the black box.

Following the aphorism of an old Italian politician, "there are no simple solutions to difficult problems", I'm deeply distrustful of the first goal (explaining AI) when we involve sophisticated AI tools. It seems to me to be the umpteenth expression of the simplification conspiracy that today is quite strong, in conjunction with other movements such as No-Vax, flat-hearthists, and so on. Facing a difficult problem by requesting a solution so simple it can be discussed among friends over a beer, and if the solution appears difficult to say it's wrong (possibly fake), may comply with the goal of improving the user experience, yet not with that of enhancing wealth of knowledge of people who ask to understand.

Simple (not trivial!) expressions on few variables, if you want me to understand.

On the other hand, interpretation as well need to be relatively simple. This entails:

1. simple expressions (for instance a linear combination of a set of polynomials)
2. a small number of parameters (say, at most 5) so that we may maintain their jointly relevance on the functions
3. a small number of variables (say, at most 5 again) for analogous reasons.

A simple example of actionable interpretations is shown in the following picture. The goal is to relate car fuel consumption (for instance expressed in Miles Per Gallon (MPG) to relevant features of the car, such as: $n°$ of Cylinders, engine Displacement, engine Horsepower, Weight, Acceleration, Model Year and Manufacturer Country.

Using a public dataset[10] collected between the years 1970 and 1982 and properly split in training and test sets, we can train a deep neural network to discover this relation. But we can also approximate it through a linear combination of

10 https://archive.ics.uci.edu/ml/datasets/auto+mpg.

univariate polynomials of degree 3 on the sole variables: {engine Horsepower, Weight, Acceleration, ModelYear}. The graphs below contrast the efficiency of the two learning tool by a scatter plot of points pairing the observed MPG with the values computed respectively by the neural network and the mentioned polynomial system that we call Marginal Functions Ensemble (MFE). The straight line $y = x$ in blue is the site where these pairs would fall in case of perfect learning (hence with the computed value equal to the target value). We see that in the training phase the two method are almost equivalent, whereas in the recall phase MFE outperforms DNN, as a consequence of the less number of parameters to be identified, thus avoiding over-fitting[11]

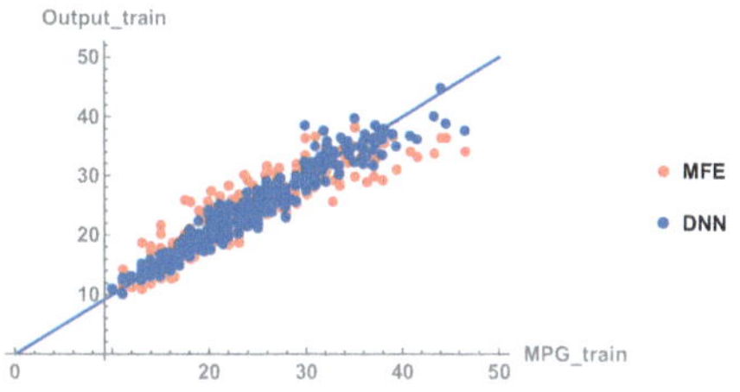

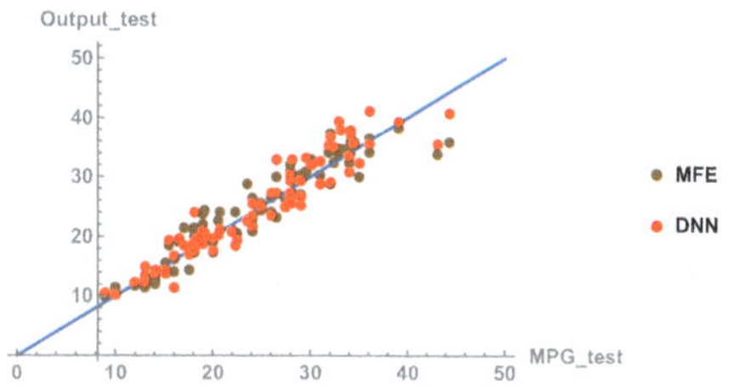

In greater detail, the effects of the four variables linearly combine with coefficients {0.388298, 0.586253, −0.224216, 0.634796}, where we see that the ModelYear has a greater relevance than Weight and Horsepower. Obviously the simple marginal dependency on the MPG by the single parameters shown in the picture below are not sufficient to understand the consumption data. Indeed, these data are spanned along a time period where the technology improvements produced a greater efficiency of the engines in general.

[11] As mentioned earlier, a common trap of the supervised learning procedure is that the cognitive system learned to reproduce exactly the training set, without understanding the underlying function linking the inputs to the outputs (again, recall the early task "repeat with your own words"). This drawback is named *over-fitting*; it may occur, for instance, with neural networks when the number of free parameters is very large and the number of training cycles is large as well, so that the training set tends to be encoded in the connection weights and biases exactly as it is. We normally recognize this phenomenon when the output error on the test set notably increases with the decrease of the error on the training set (see the picture below, where n_hid is the large number of hidden neurons employed).

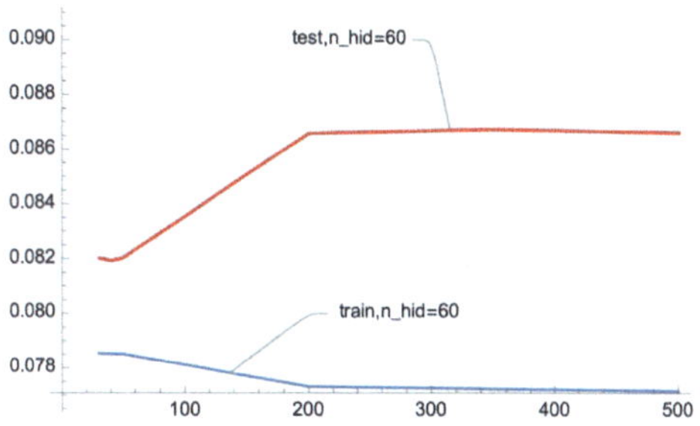

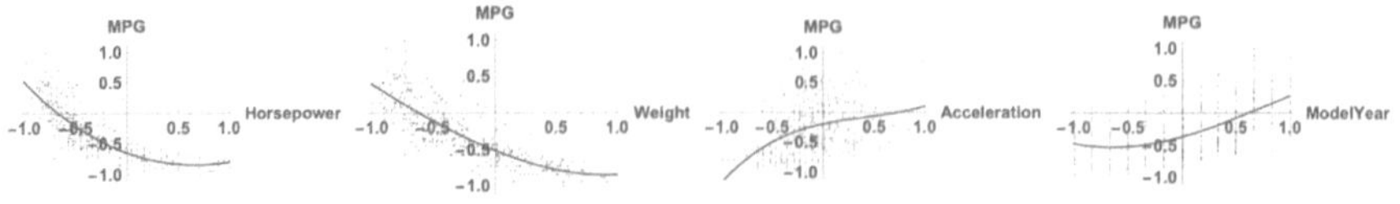

This generates some chimeric effects. Namely, the greater efficiency of the engines with time overwhelms the negative effects of the weight and horsepower on the consumption, since these quantities decrease with time. Vice versa, the negative coefficient of the acceleration takes into proper consideration the negative effect of the acceleration, even in the presence of more efficient engines, so unveiling the chimeric effect of the marginal dependence of MPG on this variable, due to the contemporary decrease of weight and horsepower with the increase of acceleration, in turn increasing with time.

5.2.1 Understanding Is Really a Social Revolution

But who wants out-of-the-box solutions? This is really a dramatic question. The standard approach to the analysis of data like the lately discussed ones, at least in the framework familiar to me, is:

A minimalist approach to solutions,

1. Sniff on the WEB *last miles* applications that may be implemented in our analysis
2. if any application requires more than one hour for its implementation, discard it (we have no time to waste, time is costly)
3. run the remaining applications, possibly with some wisdom/patch to override empty solutions, overflows and other flaws.
4. take some elementary statistic, such as average or majority, to synthesize the results of the different applications.
5. document the APIs and the entire procedure

and a challenging one.

An out of box approach requires the users to devote great effort to:

1. delving into the problem whose solution they are called upon to provide,
2. searching for a solution that better exploits the available information (exactly because they have understood the problem in depth),
3. taking their own responsibility for the effects of the solution decided upon,
4. implementing the solution procedure and
5. validating the procedure.

Items 1, 4 and 5 of the last list may clash with some mental laziness induced by the world as a service paradigm. Item 2 could prove the most appealing as it challenges the ability of the users, with obvious rewards in case of success, and the most effective, since it may introduce profitable novelties in the current operational framework. By converse, item 3 may be the most unappealing, since it involves responsibilities and anxieties in working environments that prefer to feel guaranteed.[12] Thus, as mentioned in the abstract, the selection between monolithic and out-of-the-box approaches depends on the targets of the stakeholders. Rather than a fight between two parties, where the winner overwhelms the other side, thus evoking categories of oppressors and oppressed people and other ethical qualification in the realm of the soul, we may assume that society evolves along a changing balance between costs and benefits of lists like the above two ones, as it is perceived by its members.

Actually, from a scientific perspective the search of this equilibrium has been solved at least half a century ago with the introduction of the strategies for stochastic optimization algorithms, having the Simulated Annealing [7] as a template since the eighties of the past century. Namely, being in search of a minimum, with the reference to the objective function in the last picture on Sect. 1.3, while the greedy procedure moves the ball exactly in the direction where the objective function decreases (as suggested by the standardized trend of the simplification conspiracy – the monolithic solution), if we start from the right, a procedure that allows us from time to time to make moves in the opposite direction (in search of out-of-the-box solutions) may prove more effective if it allows the ball to overcome the central hill (on whose right there exists the wanted minimum). If we modulate the possibility of the opposite move with a probability that decreases with the induced increase in the objective function, we may establish some asymptotic insurance of reaching the target minimum. Just to fix the ideas a basic instance of simulated annealing algorithm is the following.

Challenges, just when they need,

Algorithm 3 Simulated Annealing

Given the k-dimensional Boolean hypercube $\mathscr{I}^k$, i.e. the set of all binary strings $s = (b_1, \ldots, b_k)$ of length k, a loss function $L(s)$ to be minimized, and a generator of values of a Random Uniform variable U,

1. Initialize s as a sequence of k values $(s_1, \ldots, s_k)$ of $HVS(U - 0.5)$, $\hat{L} = L(s)$ and T,
2. Draw u, pick s_j so that $(j-1)/k < u \leq j/n$ and get s' by flipping the value of s_j (i.e. $s_j = if[s_j = 1, then \;\; 0, 1])$
3. Draw $u, if[u < e^{-\frac{L(s')-L(s)}{T}}, then \;\; s_j = s']$
4. Manage T
5. Go to 2

The rationale is that, if we face an NP-hard problem defined on Sect. 3.6.5, like the knapsack problem, we should in principle explore a number of candidate solutions that is exponential in the length of the input (e.g. the number of items to be filled up into the knapsack). So, to avoid incurring in the curse of the exponential function as in the picture below, on the one hand we gently perturb the candidate solution s by flipping randomly one of its binary components, then, if $L(s') \leq L(S)$ we accept this change, otherwise we accept the change only with a probability that is as much small as the jump $L(s') - l(s)$ if high. In this way, we may explore the *entire* hypercube in search of the minimum, being assured that we cross the site where L takes its overall

and for the amount they need

[12] Nothing to do with the precariousness plague of many of the current employment contracts.

Random strategy is not to simply toss a coin.

minimum with a probability that increases with the tuning parameter T. It may range from values close to infinite that ensure a great speed of the exploration but small differences as to the probability of visiting a site, the optimal one included, to values of T close to 0 that highly differentiate the latter probability, but reduce enormously the exploration speed, with the risk of remaining almost stuck in a local minimum. As such, the algorithm lacks a termination test that will stop the procedure according to some criteria (satisfying approximation, amount of runtime, etc.)

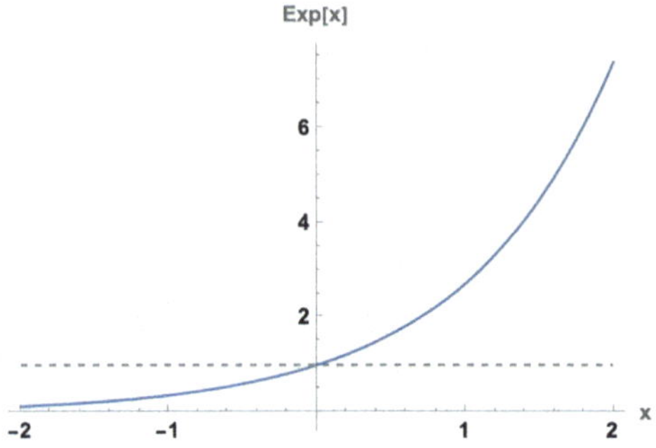

Coming back to the strategic dilemma: monolithic versus out-of-the-box, the exploration of the solution search space is exactly represented by the out of box solution. As the parameter T suggests, we need to be daring in trying new solution to span their space, yet competent as well to avoid a never ending search. But this is a science perspective. Market is another matter.

5.3 Bordering Own Knowledge Realm

Out-of-the- box solutions, even more than monolithic ones, highlight the lion problem. When Denis Diderot and coauthors carried out the great work of writing the Encyclopédie, there was the *Enlightenment* idea that sciences may be wrapped up into a simply ordered structure, as shown in the following picture. The idea was that their unified understanding will "empower human will and thus help to shape the social issues[13] that the French Revolution would address."[14]

[13] "What characterizes the philosopher and what distinguishes him from the populace, is that he does not admit anything without proof, that he has no acquiescence at all towards fallacious notions and that he precisely establishes the limits of the certain, of the probable and of the doubt. This work will certainly, in time, produce a revolution in minds and I hope that tyrants, oppressors, fanatics and intolerants will not benefit from it. We will have done humanity a service"; private correspondence of Diderot with Sophie Volland on 1762.

[14] André-François Le Breton, Jean-Michel Papillon, Ephraim Chambers. Encyclopédie, ou Dictionnaire universel des arts et des sciences. 1745.

src 31

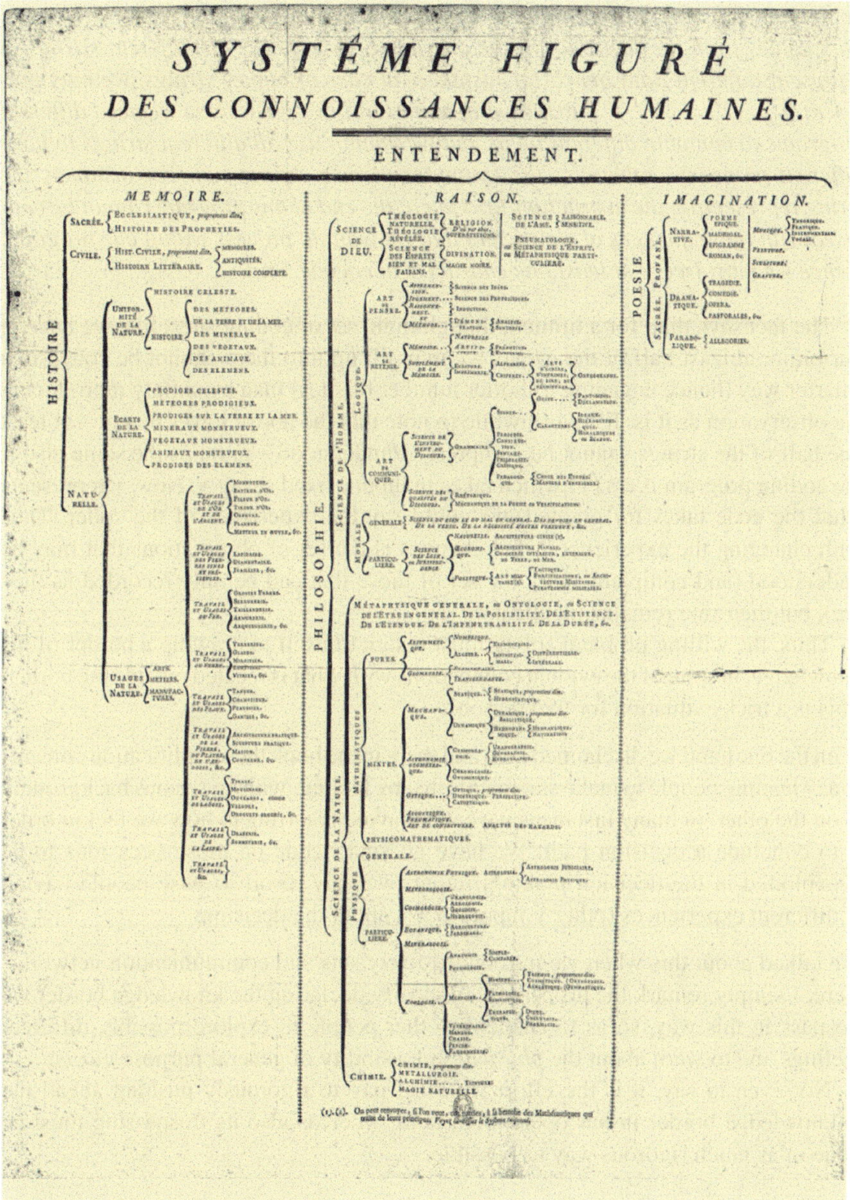

Today the unified framework and the overall container become unfeasible. It's a matter not only of the volume of the knowledge to be gathered, but mainly of the complexity of their taxonomy, which cannot be represented through a DAG (direct acyclic graph) and the continuous updating of the contents. Let's recall the hypertext organization of Wikipedia and mom's advice "not trust know-it-alls," which is formally supported by the following fact.

a lower bound from Kolmogorov complexity

Fact 1 *At least one half of the data items represented by all strings of characters up to a given length have a Kolmogorov complexity*[15] *higher than their own length. This comes from the fact that on the one hand we may use all different strings to represent different data items (for instance, in case of binary strings, we may represent* $\sum_{i=0}^{n} 2^i = 2^{n+1} - 1$ *items using at most n bits).* Vice versa *we need different programs to compute different items, but we cannot use all different strings to have different programs, otherwise the computer we use cannot run. For instance, the string representing the instruction* `End` *put at the end of any program cannot be confused with the sequences of characters that we want to process within our program. Hence we must freeze at least one character to encode* `End`*.*

The fact says that, for a limited, though high, set of coding sites, like we have in our brain, at least half of the events we may fill up into them cannot be coded in a smarter way (hence engaging a shorter number of sites) than by simply reproducing the observation as it is. It's worthwhile to note that the above lower bound – at least one half of the strings cannot be compressed/understood – remains the same also if the coding program is a $\pi(y)$ which takes in input a fixed string y. Now, y represents what the code takes for granted, we could say the experience of the coder. Thus with changing the experience, what changes is the set of observations that may be understood (and complementary the set of those that can be only recorded as they are), but their rate remains identical.

Thus, the willing problem solver must resign himself to drawing a border of his knowledge, in spite of the evident connections with what is located outside the border. This is a tricky situation for two reasons:

- on the one hand we disclaimed here and there in the book the simplification conspiracy leading people to make statements just by feeling, with no serious background,
- on the other, in many instances we cannot avoid resorting to how we feel in order to conclude a decision path. We have no smart code on those occasions to be exploited in the decision making; rather, we may resort to wise people having different experiences (other y input) to ask about the decision.

We talked about this when we introduced fuzzy sets and communication networks. Here, I simply remark the properness of clearly declaring the knowledge border we trespass in this way, so as to challenge other people to explore possibly different feelings and to warn about the *physical* unfeasibility of general purpose *gurus*.

Not even to say, it is the effort we must pay to rigorously pushing ahead the unknowledge border in our operational field, whereas also its trespassing must be done in as much rigorous way as possible.

[15] Recall Kolmogorov complexity introduced on Sect. 4.7.1.2.

5.4 No Thanks, You Must Repeat, Please!

Be patient, a few real or conceptual exercises to fix the ideas.

This is the punch line of a joke where a tourist receives a very intriguing list of instructions to reach a famous square in a town. Something like: "go straight for 200 m, then at the newspaper stand turn right; after 42 m you find an equestrian statue, go in the direction of the horse's tail for another 33 m; then you find . . etc."

In place of saying "no thanks, you must repeat!" like the sadistic interlocutor tells the tourist in response to his thanks, on the following pages I propose to the reader a small set of exercises that would stress the topics of this book in a pragmatic way. Obviously the reader may find solutions that are different from those suggested in the book. Simply, they would denote a different compatible world, thus paradoxically confirming the leitmotif of the book.

Exercise 5.1 Operation Ancient Babylon. In 2001, a huge army was sent to Iraq with the contribtion of my country as well. To link the notions of my bcd course on Probability and Statistics to reality, my colleagues and I proposed to our students a test more or less like this (from memory). The Department of War has a fixed budget to assemble troups for a war mission; thus, its goal is to minimize the economic risk R due to the death of the nation's own soldiers. The variables coming into the play are:

1. the cost of training and equipping a soldier c_e
2. the probability of a soldier surviving p_s

Assume that the relation between p_s and c_e is

$$c_e = \frac{3}{1 - p_s} - 4p_s - 2$$

- what is the most convenient p_s?

The first picture in the figure below shows the trend of c_e to be almost constant up till $p_s = 0.2$ then it rapidly increases. This increase makes the economic risk $R = (1 - p_s)(\frac{3}{1-p_s} - 4p_s - 2) = 4p_s^2 - 2p_s + 1$ to have a minimum on $p_s = 0.25$, since too high costs would improve this probability so as to reduce the first factor in R. By the way, a short sensitivity analysis reported in the table shows how they change the most convenient p_s, denoted as "srv prob", related to soldier cost, denoted as "mincost", and overall risk, denoted as "risk", if we modify the training cost by an exponent "exp" to which to raise the original cost model.

COST	RISK	SENSITIVITY
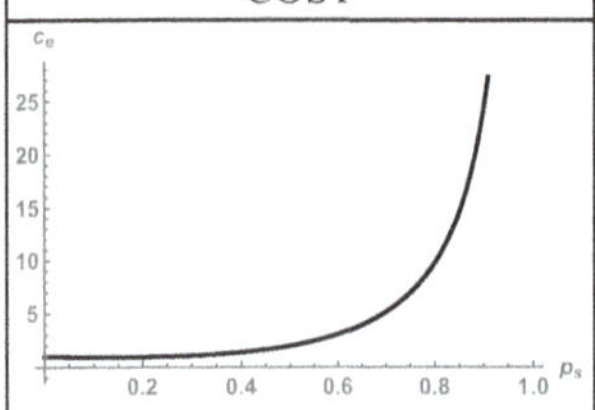	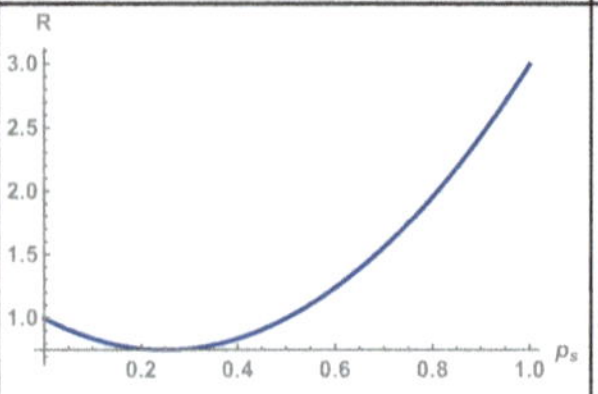	see table below

exp	0.5	1.	2.
srv prob	0.546	0.25	0.191
mincost	1.558	1.	0.891
risk	0.707	0.75	0.72

The table simply says that if the trend of the training cost grows more rapidly with p_s it is more convenient to establish a low p_s to optimize the risk. When we delivered this test some social stakeholders expressed disappointment, since the exercise appear to sabotage the national war effort. That was not the intent. The cost function was absolutely invented to get the risk function to show a minimum. The surviving probability 0.25 was definitely far from the reality of war, fortunately. My colleagues and myself just wanted to highlight the relevance of the economic aspects of that venture, and any similar one. In the economy of this book, I bring up this exercise to enhance the role of the objective function in the management of social actions. It's an argument that we can extended to analogous problems, such as:

- **The like fabric** Although the problem must be enunciated in a more complex way, the target of marketing companies is to make money, as universally accepted. A main way of achieving this is to sell the products at a planetary scale. A main strategy is to endow them with a pleasant user experience. Let's schematize the problem by reducing the user experience to a Boolean variable (good→ 1, bad → 0), so that p_s of the above exercise refer to the probability of a good experience, and consider any cost of the production as a function depending on p_s; a rudimentary marketing goal could be to minimize the risk of getting back refused products. Actually, the true marketing dynamics are more complex. However, even the above risk shows the need for a trade off between production costs and product quality, the latter as a primer of the user experience. In any case, as in the matter of equipping soldiers, we may consider different cost functions. For instance, the investments in a pervasive advertising campaign may prove lower than those for a concrete improvement of the quality of the product.[16] As a matter of fact, in spite of a relatively limited amount of budget currently devoted by companies to (both traditional and digital) advertising, around 10%, in many instances we may realize that the quality of the advertising content is higher than that of the products advertised. In sum, strictly from an economical perspective, it may prove cheaper (and mandatory as well vis-a-vis competitors) to favor marketing over product to improve the user experience and reduce the overall production costs. Moreover, recalling the early sensitivity analysis done above, higher costs per product find the trade off with the user experience positioned on lower p_s values. This may

[16] https://www.gartner.com/en/marketing.

frame the strategy of low quality factories: good advertising, poor product quality and high discard rate by user.

- **The chat box** In the same thread, this software is progressively replacing call centers to answer customer questions. Actually, the dynamics are a bit more modulated. From in-house call centers, which are rooted in the company assets and served by specifically trained people, we moved to outsource centers with general purpose/general language responders. At this point it has been evaluated that a robot could perform equally or even better, thanks to some special effects (tone of voice, background music, etc.). It's interesting to follow this path on the part of the users. Initially, they were confident about call center assistance, then they resigned themselves to poor quality answer of the general purpose responders; at that point, moving to the chat box made no big difference, with the vague reward of benefiting from top technologies.[17]

Exercise 5.2 Publish or perish.

Let me come to this case study that's quite close to my own experience.

A young Ph.D. is planning his work to get an academic position in relatviely few years. On which statistics should he rely to be successful?

Actually, answering this question is far from straightforward. As members of many academic competition committees, my colleagues and I are continually involved in long debates on the matter. The matter is so delicate that from time to time someone proposes just drawing straws to choose the winners. It's not a joke, it is resorting to the probabilistic algorithm we mentioned on Sect. 2.4.2 as the best strategy when there are no elements to take better decisions; it's a strategy frequently evoked in political instances, and that says a lot. We may say that it is a problem of goals, as the profile of ideal researcher may vary from committee to committee; but mainly it's a problem of well-behaving statistics, which we know are the key to any inference on the profile of the candidates.

But the young Ph.D. won't rely on chance. Hence:

1. he scrolls the web to find the most authoritative journals and books in his research field, then,
2. he sniffs the social networks in the area,
3. he talks with prospective colleagues (young and old).

By performing the first task, he gets materials from which to take inspiration for his work and uncover challenges for his production.

From the second he gets directions for his career. The main statement is "publish or perish". *Per se* this is a honest suggestion: the *clou* of your research activity is the statement of the results in writing, so that they may be shared with the scientific community. Then, opening a page of any scientist on Google Scholar or on Researchgate he finds three indicators:

---

[17] Not to mention the recent *metaverse* trend, where, inspired by [8], major social networks plan to introduce avatars as new appealing interfaces between users.

1. reads
2. citations
3. h-index

The first two are self-explicative: the number of times a your papers have been read, at least through the selected platform, and the number of times your papers have been cited by another scientific paper, at least according to the above platform. The third item is the fatal one. It refers to a number of scientific papers you have produced, as reckoned by the platform. However, it doesn't refer to the total number of papers, maybe ranked on the basis of the authoritativeness of the publication site. It relates this number to a sophisticated science-driven version of the Like Ranking. h-index= h for a given author says that he has published exactly h papers, which have been individually cited by at least h other papers. The algorithm is simple, clear and well sounding, so that this index represents an official stronghold of many decisions on academic careers. Intrinsic defects such as self-citation (same author, different paper), different indexing (even by a factor 2) by different platforms or trust phenomena, look like side effects.[18]

The third action by the young Ph.D. is aimed at understanding the survival strategies as they emerge from the practical experience of young colleagues and wise suggestions from the older ones. From the former he derives another must statement: "we must *make* numbers." Actually, they refer to the numbers that meet the thresholds of the competitions. In essence: "many papers", "many citations by colleagues", and "ancillary academic activities" which, according to local custom, the committees view positively. Now, if he plans to publish 10 papers per year,[19] it means that he can devote one month per paper (not counting summer and end-of-the-year holidays) for the time exceeding other activities in the normal log of a junior researcher. In particular, if he's working on computer science, he cannot plan to develop a mean complexity code, because it is too time demanding. Rather, he can adopt existing codes, when available for free, with some smart adaptation to be operated on the *skin* of the code. And this fits the URM paradigm. Moreover, if he plans to collect many citations, according to the h-index democrature, he must cultivate public relations with colleagues around the world. In principle this is a worthwhile activity which stimulates scientific cooperation and new findings, since research is definitely a social activity. The risk however is the same that we considered for factories: selling advertising in place of products. Actually, if we devote 30% of our time to routinary duties and 20% to public relations, we may share to be cited half month of research per paper.

The young Ph.D. consults old colleagues as well. Here the story becomes a bit intricate. Scientists are progressively replacing priests in the role of depository of the life's secretes. In that they inherit the esteem of their role and mistrust on many of their practices. In particular, besides gratitude for research results that the scientists grant to humanity, as for careers, the common thought is that they are highly affected by the nepotism of committees. So that academic competitions are one of the instances

[18] Jokes on the h-index of famous scientists like Einstein early in their career abound.

[19] And the instances of prolific author producing more than 100 papers per year are frequent.

mentioned above, where tossing a coin is invoked as a way to erase the nepotism. Now, we have agreed that while uniform variable U is a wonderful seed with which to feed explaining functions, knowledge about the field, the sample, is a proper way to decide the explaining function. The result is not a sharp number, but rather a tight distribution of compatible values. Knowledge is the observation of the scientific activity of the candidate, where a privileged (when faithful) observer is the team inside which the candidate has worked. Thus "nepotism" and "professional knowledge" of a candidate are fuzzy sets whose membership functions are slightly overlapped so as to be well distinguishable.

But this case study deserves another important aspect. If the goal is one paper per one or two months, we need a quick referee process, otherwise it creates a traffic jam, and a soft refereeing as well, given the time shortage to prepare the paper. These needs have given rise to the flowering of impromptu editors who promise to publish your paper on a journal in the field in 10–50 days at a price ranging fro a few dozen of Euros to one thousand Euros. There is a wide spectrum of these publishers, ranging from well-reputed brands such as IEEE and Springer, to those denoted as predatory publishers, where the paper is just an item to be sold to the maximum mutual profit of publisher and authors. In principle, all these journals adopt the peer refereeing process in accepting a paper or not. In practice, a referee may receive a letter like the one below

> We kindly invite you to review this paper and evaluate its suitability for publication in xxxxx. The article abstract is available at the end of this message.
>
> If you accept this invitation we would appreciate receiving your comments within 1 week.

After a pair of rejections the editor turns to his *internal refereeing division* to conclude the publication in the promised 10 days.

On the other hand, a candidate author may receive an invitation like this

> Dear Author, Publish your article within 48 hrs in multidisciplinary Journal.
>
> Review report within 24 hrs by the team of 250 + reviewers.
>
> Low publication fees (1700 INR/ 35 USD) including DOI.
>
> Provides certificate of publication to all authors.
>
> QUEST Journals got 9th Ranking by AQCJ (African Quality Center for Journals) - Top 10 Journals Ranking.
>
> Indexing: Index Copernicus, Cross Ref (USA), NASA ads, ANED (American National Engineering Database), Google Scholar, Open- J Gate.
>
> QUEST Journals provides DOI (Digital Object Identifier) to each article. QUEST Journals DOI is 10.35629.
>
> Papers are invited for QUEST Journals JULY 2021 Issue related to all fields of Engineering, Management, Medical & Dental Science, Pharmacy, Applied Sciences, Nursing, Humanities, Agriculture and Environment science and Social Science etc.
>
> IMPACT FACTOR: 7.21"

But the business doesn't end here. If either the candidate author or the candidate reviewer would like to understand whether the sending journal is a predatory one, he can solve the question by consulting a specialized web service, like Cabells' Predatory Reports. But also this is a business, so that he must pay a subscription to get an answer.

Before ending the exercise, let me remark that Open Access publication has great merits as for the diffusion of the papers which do not require fees to be downloaded. Thus our aim was not to demonize it. Rather, the exercise highlights the role of proper statistics as a way of exploiting fruitfully available data. Bad statistics may set up perverse systems that prove difficult to erase.

Exercise 5.3 A world beyond Gaussian Distribution This exercise was the starting point of my Approach to Algorithmic Inference. Actually, being involved in computational learning theory, it was clear to me that the problem of learning a class of boolean functions is just a problem of computing a confidence interval of the error probability (see Sect. 3.3.1.1). Thus it's a matter of the parameter of a special Bernoulli variable, whereas no clear theory was available for computing confidence intervals of the parameter of this variable. Therefore, that the problem was generally solved either in terms of lose upper and lower bounds,[20] or in the approximation of a sum S of Bernoulli variables with a Gaussian variable, and also with this approximation the theory is sloppy.

The above approximation holds when the sum applies to a large sample. Let me bring up the story of my son's 20 games again, where the approximation may result unsatisfactory. Let come to a sample of only $n = 10$ games to stress the inadequacy of the approximation. From Sect. 2.1.1.1 we know that the probability of victory after n games, k of which resulting in wins, has a distribution law bounded as:

$$1 - F_{K_p}(k) \leq F_{P_k}(p) \leq 1 - F_{K_p}(k-1)$$

which puts in relation the CDF of $S = K$ for fixed $p = p_0$ with the CDF of P for fixed $s = k$. According to the considerations on Sect. 2.1.3.2 and hiding the index in P_k per usual, we may state that

$$\mathscr{P}(p_l < P \leq p_r) = F_P(p_r) - F_P(p_l) \geq 1 - F_{K_{p_r}}(k) - \left(1 - F_{K_{p_l}}(k-1)\right)$$

where (p_l, p_r) is the desired interval for P. If we want P to fall inside it with a probability at least equal to $1 - \delta$ we may fix the extremes of the interval such that

$$1 - F_{S_{p_r}}(k) = 1 - {}^{\delta}\!/2; \quad 1 - F_{S_{p_l}}(k-1) = {}^{\delta}\!/2.$$

Solving the above equations for any value of $k \in \{0, \dots, n\}$ we obtain the typical cigar-shape confidence domain for a Bernoulli variable, like the one delimited by the green curves in the picture below. For each k the intercept of the curves with the vertical line on k identify the extremes of the $1 - \delta$ confidence region. By varying

[20] Classics bounds are obtained using Chernoff inequality [2].

δ we obtain different nested domains (the one with lower δ includes the one with higher δ).

We may obtain similar domains starting from the approximation of S distribution to a Gaussian distribution. Namely, in the classical scenario, while tests of hypotheses are very simply checked (through sufficient statistics involving binomial distribution), the usual way of computing confidence intervals passes through the Central Limit Theorem [4]. For X random variable with mean μ and standard deviation σ, denoting with $\overline{X}_m = \sum_{i=1}^{m} X_i/m$, it states the asymptotic property:

$$\lim_{m\to\infty} \mathrm{P}\left(\frac{\overline{X}_m - \mu}{\sigma/\sqrt{m}} \leq x\right) = \Phi(z),$$

where Φ is the CDF of a standard Gaussian distribution. In spite of the m instantiation definitely far from ∞ ($m = 10$), we commonly specify X_m with S_P/m, μ with p and σ with $\sqrt{p(1-p)}$ and obtain p^*_{dw} and p^*_{up} as the solution of the equations:

$$\frac{s_P/m - p^*_l}{\sqrt{p^*_l(1-p^*_l)/m}} = z_{1-\delta/2}; \qquad \frac{s_P/m - p^*_r}{\sqrt{p^*_r(1-p^*_r)/m}} = z_{\delta/2}.$$

By instantiating s_P on the various vales of k, we obtain the analogous confidence domain delimited by the red curves in the same picture.

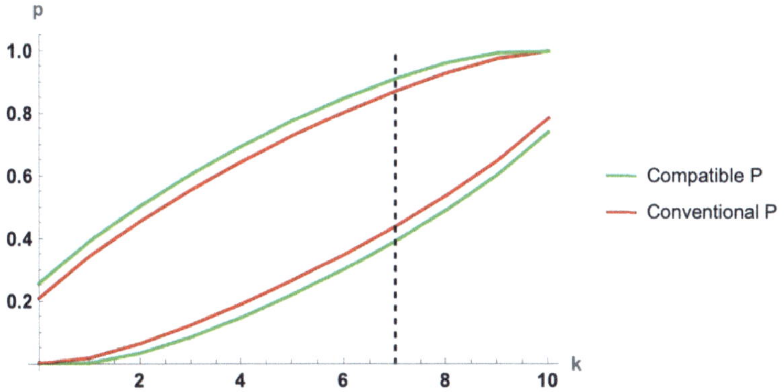

The comparison between the two domains denotes a wider extension of the confidence intervals referring to compatible P. The difference between the two domains is due to two effects:

1. the continuous approximation of the S_P/m distribution law shifts the confidence intervals upwards, conversely
2. the witnessing functionality of the $(s_P + 1)$-th point in (Metaprob) raises the upper bounds of the confidence intervals in the algorithmic inference approach.

I also recall the semantic difference between the two kind of intervals. The intercepts with the green curves, for instance with the vertical line $k = 7$, identify an interval where to find with probability $(1-\delta)$ a value of P compatible with k 1's over m bits, whatever their source (provided they constitute a random sample). This means that also gathering samples from different sources (hence with different p's) each time, our intervals ensure an overall confidence that is independent of the distribution of the sources (which, consequently, we may ignore). Therefore, if you adopt this way of computing the intervals in a long story of samples, the frequency with which the current parameter is gathered by the current interval is asymptotically equal to $1-\delta$. The acceptation of the classical confidence interval is more *plastered*. It says that if you draw many samples from the same Bernoulli distribution, hence with same p the set of $(1-\delta)$-confidence intervals you compute, for instance using the Gaussian approximation, will include p with probability $1-\delta$. Hence you decide to use this probability as a measure of confidence that p is included in the confidence interval computed on the value k you measured on your actual sample. To delve further into this point the reader may refer to [1].

By comparing (Metaprob) with (Metaprob1) we realize that what changes when we pass from the inference of P of a Bernoulli variable to that of P of the error in learning a Boolean function is an overload of k by d and 1, respectively on the right and left extreme of the inequality. Thus we have the same cigar domain, just a bit displaced in the (k, p) plane, with analogous manageability in tackling the quality of the learned Boolean function.

The previous exercise and previous consideration widen in practice the spectrum of compatible worlds we are able to manage: Not only those more or less asymptotically linked to the Gaussian distribution, but many more worlds. Actually, basic statistical problems currently deal with those worlds, for instance in problems of survival analysis. But, when things become complex, adopting the Gaussian distribution as the reference one renders solutions feasible and comfortable, even though possibly not satisfactory.

I just recall the variational autoencoder (VAE) [6] and Generative Adversarial Networks (GAN) [3] that are currently the top tools to generate distributions *similar* to the one observed through a sample. With the former the procedure starts from a reference distribution and changes its parameters in order to cope with the observed sample. With the latter we train a neural network to generate a new sample where no skilled observer (actually another neural network) may distinguish the original from the copy. Needless to say, the most common reference distribution as distribution of seeds of the target variables in both strategies is the Gaussian distribution, followed by the Bernoulli distribution for VAE and the uniform one for GAN.

The central limit commodity

As recalled in the previous page, the central limit theorem states that:

- if we gather a huge set of variables
- if we consider of them uniquely the sum as the overall effect

- if we destroy any relevant (not necessarily anyone) relation between the values that they jointly assume[a]
- then we may control the behavior of this sum through a Gaussian distribution having as the mean the sum of the means of the single variables and as a good approximation of the variance the sum of the corresponding variances.

The Gaussian distribution is an ideal distribution to describe this kind of variable: it is endowed with a central value that's *equivalent* to the spread values of the variable when, again, what is important is their sum, and it's symmetric around this value, thus weakening any aim of investigating the causes of the shifts.

Again as in the egg-chicken dilemma, it's questionable which came first

- the globalization of the business as a safe framework where the revenue of the big companies explodes and the risks are compensating thanks to a diversification of the ventures – both effects of the Central Limit theorem,
- the standardization of the requests of the customers (any of us as overall) allowing the globalization and rewarded by an abundant satisfaction of those request – enforced by the companies and well accepted by the customer to enjoy the Central Limit commodity.

Both fall in the perspective of the URM paradigm.

[a] These conditions may aptly characterize the totalitarianisms.

Exercise 5.4 In the happiness country, people are rich enough to buy ever better performing cars, which they see as a *natural* extension of their mobility capabilities. As such, cars are used in any circumstance and for any purpose, independently of the driver's attention level, and of his driving capability as overall. Thus a driver who is distressed, drunk, or simply distracted (by a mobile phone conversation, for instance) may be involved in a fatal accident. Public opinion is disturbed for these accidents, where lives are wasted. Thus a strongholds committee is convened to find a solution. After one month the committee concluded its mandate by offering two alternatives:

1. To reinforce the education of citizens as to proper and responsible driving.
2. To replace the driver with a robot that doesn't get distressed, drunk, or distracted by a mobile phone, and emulates with an *expectedly good* approximation the decision processes of humans.

Asked for a decision, what wold be your option?

Actually, the answer is not obvious. It fully involves the arguments of this book (and this chapter in particular), hence the URM perspective. From a centralized viewpoint, the solution is to remedy the drawbacks independently of the contribution of the solution addressees, hence the second alternative. The solution is on a global scale, thus implies a Gaussian world, with some local adaptivity and tail drawbacks amounting to a *limited* amount of fatal accidents. Abandoning the unfeasible zero defects hypothesis, this number is simply to be compared with that of the alternative solution. In light of current statistics on prototypes, we expect it to be not extremely small, also because the user of the car is relieved of any responsibility, thus becomes unaccustomed to dealing with driving situations when necessary. Whether or not drivers like to be substituted by an ADS (Autonomous Driving System) is another question to be faced in view of the user experience of the product car. For the moment, the ability to deliver in a near future cars with ADS systems in any near future represents a form of market leverage to advertising the technological power of automobile manufacturer. Meanwhile, ADS availability progressively tends to become an induced need, again of the Gaussian world.

From a user perspective the first alternative offers many more benefits. We increase our personal and social education. We don't miss our ability to drive a car, like we did for instance with our ability to ride a horse or to quickly consult a road map (after the pervasion of electronic navigators). We maintain/improve our habit of taking responsibility for our actions in challenging situations (even beyond driving).

As mentioned, choosing between the two alternative is not obvious (where intermediate ones do not resolve the basic dilemma). However, I assume that the alternative selected will clearly indicate the possible worlds toward which we will move in the future.

References

1. B. Apolloni, S. Bassis, Confidence about possible explanations, in *IEEE Transactions on Systems Man and Cybernetics Part B-Cybernetics* (2011), pp. 1–15
2. H. Chernoff, A measure of asymptotic efficiency for tests of a hypothesis based on the sum of observations. Ann. Math. Stat. **23**, 493–509 (1952)
3. I. Goodfellow, J. Pouget-Abadie, M. Mirza, X. Bing, D. Warde-Farley, S. Ozair, A. Courville, Y. Bengio, Generative adversarial networks. Commun. ACM **63**(11), 139–144 (2020)
4. M.G. Kendall, A. Stuart, *The Advanced Theory of Statistics, Volume 2: Inference and Relationship* (Charles Griffin, London, 1961)
5. N. Klein, *No logo: Taking aim at the brand bullies* (Knopf Canada, Canada, 2000)
6. M..J.. Kusner, B. Paige, J.M. Hernández-Lobato, Grammar variational autoencoder, in *Proceedings of the 34th International Conference on Machine Learning*, ed. by D. Precup, Y.W. Teh. Proceedings of Machine Learning Research, vol. 70 (PMLR, 2017), pp. 1945–1954
7. C.D.S. Kirkpatrick, Jr. Gelatt, M.P. Vecchi, Optimization by simulated annealing. Science **220**(4598), 671–680 (1983)
8. N. Stephenson, *Snow Crash* (Spectra, 1992)

Chapter 6
Conclusions

"And so what?" the tenacious reader who comes to this page in the book may ask. "Did I reach my goal" is the companion question on my part.

Actually, this book arises from a need rather than with an aim. If communication is the red label of our society and the technological leverage of its progress, sharing ideas may represent a pebble of this progress. I attempted to do it in a proper way precisely within the borderland of what is understandable and what is not. Once mystic interventions of God willings, soul and premonitions have been cast aside, the challenge lies in shifting this border a bit forward via a proper management of available information and of its lack as well.

My own thought is that, facing the huge amount of raw data, events, movies, links included, that are provided us by media, we need to develop mental tools to process them, in the part of our interest, without resign ourselves to simply labelling them with likes. Obviously, we can not be masters of all trades. Rather, we should be aware of the degree of expertise we enjoy on the current fields, and handle our challenges accordingly.

The alternative option is to conform to mainstream thought, as an extreme sclerotization of our synapses. Again, bad thing is to pretend of explain the entire world with a unifying, possibly rudimentary, paradigm (genome+synapses in this case). However, the leveling of ideas as the dark side of the current flood of biased communication, and the brilliance of young minds not yet bombarded by it, are both common experiences in today's life. It's the release 4.0 of Plato allegory of the cave[1]: from

[1] Plato. The Republic Book VII. Human beings sit in a cave in chains, with their backs to the entrance. The shadows of things moving outside are projected by the light onto an inner wall of the cave. As the prisoners have never been outside the cave since birth, they believe these shadows are reality. One of them succeeded in escaping the prison and walked outside into the light, so that he sow the real world. He returned to the cave to communicate it to the others, but they did not trust him.

B. Apolloni, *A Few Things I Know About Her*, Intelligent Systems Reference Library 219, https://doi.org/10.1007/978-3-030-94379-0_6

prisoners trusting in shadows, to divan mates enjoying fictions, to managers obeying cloud services.

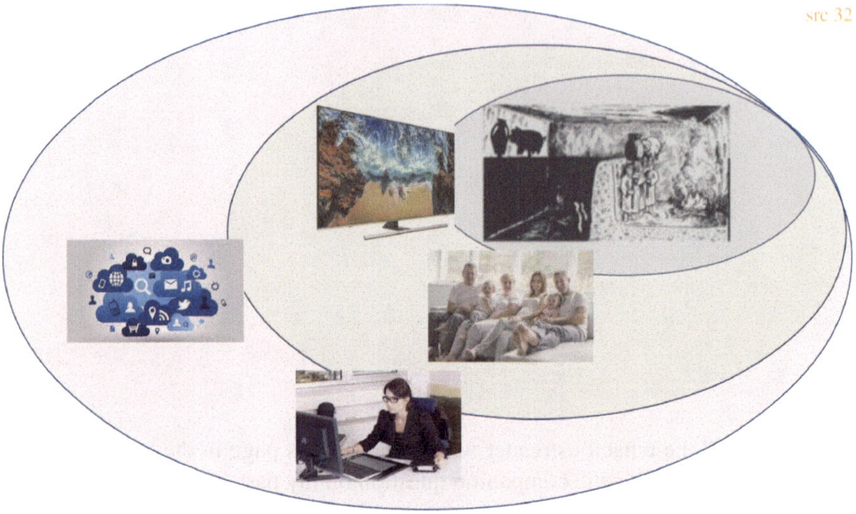

A second consideration against the mainstream leveling concerns the first component of the golden kit, the genome. As mentioned earlier, the human genome evolves via natural selection, and mankind does the same. It means that compositions of *weak* genes inside a genome render the human host's probability of surviving in the environment very low, so that their reproduction probability is small as well, and those gene compositions tend to disappear. Is this mechanism still operating? If we see the map below of the cancer mortality index around the world, we see that it represents the first premature (age under 70 years) cause in the most socially and economically developed country with an average death rate with respect overall deaths around 10% (the numbers of countries represented in each ranking group are included in the legend.) [2]. It says that the selection happens by technology, in reverse way, rather than by genes. The current pandemic as well seems to go in this direction – here underdeveloped countries cannot access vaccines and treatments – with the additional feature that it generally kills people definitely after their reproduction age.

As expected, the matter is of great interest and is variously dealt with. Besides minor genetic changes due to the racial promiscuity[2] or environmental conditions (such as transportation facilities which don't favor longer speedy runners[3]), the question remains as it was posed by some reputed researchers, such as Steve Jones of

[2] No misunderstanding, please! Though the term "race" now have different connotations – often linked to the negative social phenomenon of "racism"– races, as categories of phenotypical traits and related genes, do exist and their variety is a patrimony of humanity, which does not imply inferiority or superiority classifications, rather, a richness of exploitation paths.

[3] See, for instance, Scientific American, November 2012 (https://www.scientificamerican.com/article/what-may-become-of-homo-sapiens/).

University College London[4] who titled his famous lecture "Is Evolution Over?". There, he claimed that *At least in the developed world, almost everyone has the opportunity to reach reproductive age, and the poor and rich have an equal chance of having children*, so envisioning a *memetic* evolution that overrides the genetic one. Memetics stands for evolution of ideas as a pure epigenetic evolution. If we link it to the mind-leveling mainstream evoked in the new cave allegory, the emerging scenario is that of a standardized human kind, with no gene varieties priming new deals and the prevalence of very few stakeholders strictly pursuing their own interests. The whole could not result in an appealing scenario.

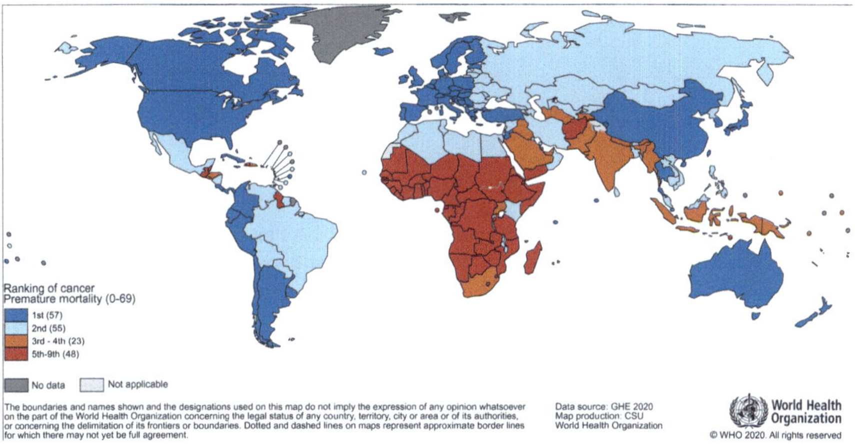

Obviously, on the negative side, criticizing to the current system is relatively easy. We may relate it to the attitude of our brain's neural network to learn from errors. Thus criticism is, anyhow, the favorite dialectical tool of many politicians, but also a fine analytical tool for philosophers, scientists, journalists, all the way to our children, like lovely Mafalda *sister* in the picture below [1]. Definitely bad things occur when it is supported by ethical categories of good and evil, possibly anchored to a divine law. Taking a positive approach is definitely harder. The aim of this book is to discuss some approaches and ideas which, far from constituting an exhaustive framework or a unified theory, provide some building blocks of a scientific nature to promote compatible worlds based on observed data and on the computational functionalities of our brains, where proactive social practices come as for corollaries.

---

[4] https://royalsociety.org/people/steve-jones-11710/.

References

1. Joaquin Salvador (Quino) Lavado, *Mafalda&Friends* (De LA Flor S.R.L. Ediciones, 2008)
2. R.L. Siegel, K.D. Miller, H.E. Fuchs, A. Jemal, Cancer statistics, 2021. CA: A Cancer J. Clin. **71**(1), 7–33 (2021)

The manufacturer's authorised representative in the EU is Springer Nature Customer Service Centre GmbH, Europaplatz 3, 69115 Heidelberg, Germany. If you have any concerns regarding our products, please contact ProductSafety@springernature.com

Printed and bound by CPI Group (UK) Ltd, Croydon, CR0 4YY
06/10/2026
02225252-0003